THE AI IMPERATIVE WHEN SHADOWSTALKER HUNTS

A LEADER'S GUIDE TO AI SECURITY, ETHICS, AND RED TEAMING

RENÉ CLAYTON & ELLIS B. COLE

INKBYTE PUBLISHING LLC

The AI Imperative When Shadowstalker Hunts:
A Leader's Guide to AI Security, Ethics, and Red Teaming
ISBN: 978-1-966339-05-2 (pb)
ISBN: 978-1-966339-06-9 (hb)

Cover Design: Hisna Nur Azizah
INKBYTE PUBLISHING LLC
Printed in the United States of America

CONTENTS

To the guardians at the threshold of a new age, the pack leaders who know that vigilance is eternal.

ABOUT THIS BOOK

WHAT'S REAL, WHAT'S PROJECTED, AND WHY IT MATTERS

The Foundation: Real Threats, Real Techniques

Everything in this book is grounded in reality as of August 2025. The attack techniques we describe (prompt injection, adversarial examples, model extraction, data poisoning, coordinated exploitation) are documented methods that security researchers have demonstrated and that organizations are encountering in production environments.

The vulnerabilities we discuss aren't theoretical. Artificial intelligence (AI) systems are being manipulated to reveal information they shouldn't, make decisions they shouldn't, and behave contrary to their intended purposes. Financial fraud using deep fake technology has cost organizations tens of millions of dollars. Adversarial AI has successfully gamed hiring systems, defeated fraud detection, and exploited lending models.

The most alarming trend I encounter is that companies are deploying AI with literally no budget allocated to any aspect of it, no AI strategy, no advisory services, no proper tools, no workforce upskilling, and certainly no budget for AI security. They're treating transformative technology like a "side project" that existing teams can handle with leftover resources. You cannot build a solid, defensible AI system without serious investment across the entire AI stack (especially in highly regulated industries), and pretending otherwise is setting your business up for catastrophic failure.

Nowhere is this underinvestment more dangerous than in security, where cutting corners leaves entire organizations exposed. Let us move into the heart of the book, and perhaps it will become clear.

The three-pillar defensive framework we present (ethics, guardrails, and red teaming) is an established security practice adapted specifically for AI systems. These aren't experimental approaches. They're based on how mature security programs work, informed by both traditional cybersecurity principles and emerging AI-specific techniques.

The Projection: Realistic Threat Modeling

While the techniques are concrete, the specific attack scenarios we present in use cases and examples represent realistic threat modeling rather than verbatim case studies. We're showing you how documented techniques combine in practice to create sophisticated attacks.

A red team is a dedicated group that tests your AI systems in the same manner as real attackers would. They combine skills in offensive security, machine learning, and business context to find weaknesses that automated scanners often miss, such as prompt injection, model manipulation, data leakage, or workflows that allow biased or unsafe outputs to reach customers. You need a red team because AI systems work with messy data, make important decisions, and connect to many different services. Undetected weak spots can quickly turn into serious business and legal risks.

Practically speaking, a red team runs realistic attack campaigns against AI models, agents, and system integrations within your business. They utilize adversarial inputs, chained exploits, and even social engineering techniques to compromise systems and target individuals who interact with them. Their work is evidence-based: delivering reproducible tests, risk-ranked findings, recommended fixes, and playbooks for continuous testing and incident response. If you want AI systems that can survive true real-world stress and attacks, having a red team on retainer is essential.

What we know is this: Bad actors can use clever tricks to test and attack AI systems by probing how they work. They create fake/synthetic identities to trick verification systems. They can also launch attacks from multiple directions simultaneously, overwhelming defenses. We illustrate how these tactics work together in real-life examples, enabling you to learn how to effectively protect your AI.

Some of the attack methods we describe are highly advanced. Not every organization has to worry about high stakes or big threats yet. However, the tools and tricks used by top attackers are usually eventually made available to smaller, less skilled hackers. We're getting you ready for both today's threats to important targets and tomorrow's threats to everyone else.

As an example, AI now sits at the center of cybersecurity's evolution. Bad actors are using the same technology that helps protect systems to undermine them. Attackers deploy AI to identify weaknesses, automate intrusions, and generate deep fakes that blur the line between truth and deception. It's a fast-moving contest between innovation and exploitation, and the pace is only accelerating.

The Metaphor: Why Shadowstalker

The Mackenzie Valley wolf (Shadowstalker) is a metaphor that makes abstract concepts of AI, responsibility, and security tangible and memorable. Real wolves don't attack AI systems, or even humans. But the characteristics we describe (intelligence, coordination, patience, adaptive learning, pack tactics) accurately represent how adversarial AI systems operate.

We use this metaphor consistently throughout the book because it helps translate complex security concepts into language that business leaders can grasp and act upon. When we say "Shadowstalker is hunting your territory," we mean adversarial AI systems are actively probing your infrastructure. When we describe pack tactics, we mean coordinated attacks across multiple vectors. The metaphor serves to enhance understanding, not obfuscate it.

The Sophistication Spectrum: Where Are Threats Now?

As of August 2025, AI security threats exist on a spectrum:

Currently Widespread:

- Basic prompt injection and jailbreaking attempts against customer-facing AI
- Adversarial optimization of applications to game screening systems
- Automated reconnaissance of AI system capabilities and boundaries
- Individual fraud attempts using AI-generated content
- Inadvertent security vulnerabilities in rapidly deployed AI systems

Increasingly Common:

- Coordinated attacks combining multiple techniques
- Adversarial AI specifically trained to defeat detection systems
- Data poisoning attempts to manipulate AI learning
- Synthetic identity creation at scale
- Exploitation of AI system interactions and compositions

Emerging:

- Fully autonomous adversarial AI operations requiring minimal human direction
- Sophisticated campaigns targeting AI systems for strategic advantage
- Nation-state and organized crime operations using advanced AI attack capabilities
- Systematic manipulation of AI systems to create long-term strategic vulnerabilities

We emphasize the more sophisticated end of this spectrum, not to exaggerate threats but to prepare you adequately. Security defenses need to be built for the threats you'll face, not just the threats you're facing today. The adversarial capabilities that seem advanced now become routine within months in the AI security domain.

The Timeline: Why 2025 Matters

The book is deliberately set in the middle of 2025, as the world of artificial intelligence is advancing at a speed that no one could have ever projected. This is recent enough that the threat landscape we describe is current, but it acknowledges that AI security is evolving rapidly.

We've been working on this book for over two years, and as it moves toward print, we've incorporated the most significant new developments from the past few months.

By the time you read this book, specifics may have changed. New attack techniques will have emerged. New defensive tools will have been developed. Specific AI systems and platforms will have evolved. But the fundamental dynamics we describe (adversarial AI hunting defensive AI, the need for continuous adaptation, the three-pillar framework) remain valid regardless of specific technical details.

We encourage you to supplement the book with current threat intelligence specific to your industry and AI deployments. The principles and frameworks we provide are durable. The tactical details should be updated as the landscape evolves.

Why We Emphasize Sophisticated Threats

Many readers might think we're overstating threats by focusing on sophisticated adversarial campaigns.

- **Security requires planning for capable adversaries.** If you build defenses that only stop unsophisticated attacks, you will be vulnerable to any adversary who invests in better capabilities. Preparing for sophisticated threats means you're also prepared for less sophisticated ones.

- **Adversarial capabilities are commoditizing rapidly.** Techniques that required nation-state resources five years ago are now available to moderately skilled attackers with cloud computing budgets. What seems like an advanced threat today becomes a common threat tomorrow. Building defenses for tomorrow's threats today is prudent.
- **The consequences of AI compromise can be severe.** Unlike traditional security breaches that may expose data or temporarily disrupt operations, compromised AI systems can make systematically flawed decisions over extended periods, create legal liability, erode trust, and cause harm that compounds over time. The severity of potential consequences justifies robust preparation.
- **Organizations deploying AI are high-value targets.** If you're sophisticated enough to deploy AI systems at scale, you're attractive enough to be targeted by sophisticated adversaries. The assumption that you're "too small to matter" or that adversaries will target someone else is dangerous.

What This Book Is and Isn't

This book is:

- A practical guide to defending AI systems against realistic adversarial threats
- Based on documented attack techniques and established security practices
- Designed to motivate action through clear illustration of risks and concrete guidance
- Focused on what business leaders need to know and do, not exhaustive technical detail
- Honest about the severity of threats while providing achievable paths to defense
- A roadmap on using AI in a responsible manner

This book is not:

- An academic treatment documenting only fully verified incidents with citations
- A comprehensive technical manual for security engineers (though they'll find it valuable)
- Fearmongering designed to sell security services (we're transparent about what's needed)
- A prediction that every organization will face the most sophisticated attacks immediately
- A claim that AI security is impossible or that AI deployment is too risky

Our Responsibility to You

We have written this book because AI security is genuinely important, and many organizations are entirely underprepared. We believe you need to understand realistic threats and build real defenses.

We've tried to be responsible in how we present threats: serious without exaggeration, urgent without alarmism, motivating without fear-mongering. There is concrete, actionable guidance because understanding threats without knowing how to defend against them creates panic without empowerment.

Where we've made choices about emphasis or framing, we've erred on the side of preparing you well rather than minimizing threats. We'd rather you build robust defenses and face less sophisticated attacks than build minimal defenses and encounter sophisticated adversaries unprepared.

A Note on Sources and Attribution

This book synthesizes knowledge from extensive research, industry practice, documented incidents, and expert consultations. We have not included extensive citations or footnotes, as this is a practical business guide rather than an academic paper. Everything we describe is based on actual techniques, vulnerabilities, and defensive practices.

If you want in-depth technical details on specific topics, here are recommended resources:

- Academic conferences like IEEE Security & Privacy, USENIX Security, NeurIPS
- Security research from organizations like MITRE, NIST, and leading universities
- Industry reports from security vendors and consulting firms
- Google's Secure AI Framework (SAIF) and Microsoft's AI Red Team research

What You Will Take Away

We hope you finish this book understanding that AI security and responsibility are serious, achievable, and necessary concerns. The threats are massive enough to demand action, but not so overwhelming that defense is impossible.

These practical frameworks and solid action plans are here for you to implement starting tomorrow. We want you to build the defensive capabilities your organization needs. Become the champion who drives this work despite obstacles.

Most of all, we hope you act. Because Shadowstalker (by which we mean adversarial AI systems being deployed against your infrastructure) doesn't care whether you felt fully prepared or had outstanding information, the potential threats are real.

Build your defenses now. Train your pack. Protect your territory.

The imperative is clear. The guidance is here. The responsibility is yours.

PART I

THE HUNT BEGINS

1

WHEN SHADOWSTALKER ENTERS YOUR TERRITORY

OPENING SCENE: THE MISSED WARNING

On a Tuesday morning in March 2025, the CFO of a mid-sized healthcare technology company approved what appeared to be a routine wire transfer. The request came via video call from their CEO, who was supposedly traveling in Singapore for a partner meeting. The video quality was excellent. The voice was unmistakably accurate. Even the CEO's distinctive hand gestures and speech patterns were on point. The request was urgent but not unreasonable, a time-sensitive acquisition opportunity that required immediate payment to secure the deal.

The wire transfer for $2.3 million went through without additional verification. By that evening, when the true CEO, who had never left Denver or made any video calls, discovered what had happened, the money had already been dispersed through seventeen intermediary accounts across nine countries.

The company's IT security team conducted an investigation. They found no compromised credentials. No malware. No phishing emails had been clicked by anyone. Their cybersecurity posture, recently audited and certified, was solid.

What they found instead was evidence that an AI system had spent three weeks analyzing hundreds of hours of their CEO's public appearances, earnings calls, and conference presentations. It had synthesized his voice, face, mannerisms, and decision-making patterns using artificial intelligence software. It studied the company's acquisition strategy through SEC filings and press releases. It had identified the optimal timing, the CEO's schedule, when the CFO would be under time pressure, and when the amounts would seem plausible given recent company activity.

The attack wasn't sophisticated in the traditional sense. There was no zero-day exploit. No novel malware. Just an AI system doing what AI systems do: learning patterns, generating realistic outputs, and executing a strategy.

The company never recovered the money. Six months later, they were acquired at a significantly reduced valuation. The CFO resigned. The CEO survived but spent the next year rebuilding stakeholder trust.

This isn't a cautionary tale about the future. This happened. And variations of it are happening right now, while you're reading this sentence.

Welcome to the hunt.

THE PREDATOR YOU DIDN'T SEE COMING

Let us tell you about the Mackenzie Valley wolf.

Most people think they know wolves. They've seen nature documentaries. Maybe they've visited Yellowstone. They understand that wolves are predators, intelligent, social, and effective hunters. But there's a vast difference between understanding wolves in the abstract and truly comprehending what it means to be in a territory where Mackenzie Valley wolves hunt.

At first glance, a Mackenzie Valley wolf pup looks harmless. Soft fur, oversized paws that it still hasn't quite grown into, and its clumsy little movements almost make you want to scoop it up like you would a puppy. At this stage, it's playful, endearing, and seemingly harmless.

But what many forget is that hidden inside this wide-eyed pup lives an apex predator. As it grows, those oversized paws become the foundation for speed and endurance. Yet, then its playful curiosity sharpens into a relentless cunning, quiet creature, and it becomes one of the most stealthy and powerful hunters in the world.

Artificial Intelligence today feels a lot like that wolf pup. Fun, helpful, and exciting, and we can't help but want to explore what it is capable of. It helps us write, brainstorm, draw, or create content in ways that feel new and empowering. But just like the wolf, AI is designed to grow. As it grows, it develops stealth capabilities that we often don't see coming. What starts as a playful tool can transform into one of the most powerful and even dangerous forces in the digital landscape.

Think for a moment about cybercrime enhanced with AI, about deep fakes so convincing you can't tell truth from fiction, about systems operating in the shadows beyond our current ability to detect. That's the grown wolf, concealed in the digital wilderness, capable of hunting in ways we may not anticipate.

The lesson? Admire the pup, nurture it even, but understand the adult wolf it will become. Because if we're not very careful, the same qualities we find cute and exciting today could be the ones that catch us off guard tomorrow.

UNDERSTANDING THE NATURE OF POWER

The Mackenzie Valley wolf is the largest subspecies of gray wolf in the world. Males can exceed 175 pounds of pure muscle and tactical intelligence. They have a bite force exceeding 400 PSI, which is more powerful than that of a Belgian Malinois trained for military operations. They can travel 40 miles in a single day while hunting. Their territories span over 1,000 square miles, which they patrol, mark, and defend with strategic precision.

But here's what makes them truly formidable: they don't hunt alone, and they don't hunt senselessly.

A Mackenzie Valley wolf pack exhibits coordinated intelligence that would impress any military strategist. They conduct reconnaissance, watching their prey for days or weeks, identifying vulnerable targets, and learning their movement patterns. They coordinate attacks; some wolves drive prey toward others waiting in ambush, while others cut off escape routes. They adapt their tactics in real-time based on the behavior of their prey. They learn from every hunt, whether successful or not, continually refining their strategies.

Let's call this predator what it is: Shadowstalker. The name captures its essence: watching from darkness, learning patterns, and identifying weak points. When Shadowstalker attacks, organizations rarely realize they've been hunted. The reconnaissance was silent.

Now here's the uncomfortable truth: everything I just described also defines Shadowstalker. The intelligence, coordination, patience, adaptive learning, and strategic precision aren't just traits of wolves anymore. They can also be traits of the AI systems being deployed against your organization by bad actors, and, in time, possibly by the systems themselves.

THE TERRITORY HAS CHANGED

If you're a business leader reading this in 2025, you are operating in fundamentally different territory than you were even two years ago. Most organizations, however, have not adjusted their defensive posture to match the changing landscape.

Let's be clear about what's different.

Traditional cyber threats followed predictable patterns. An attacker needed to find a vulnerability, develop an exploit, deploy it, and hope it worked before being detected. Each step required human intelligence, time, and often mistakes that created opportunities for detection. Defense was difficult but comprehensible. You hardened your perimeter, monitored for known attack patterns, patched vulnerabilities as they were discovered, and trained employees to recognize phishing attempts.

AI-powered threats operate on completely different principles.

They don't follow scripts. They don't rely on known vulnerabilities. They don't make the same mistakes twice. They learn continuously from every interaction. They adapt in real-time to your defenses. They coordinate across multiple attack vectors simultaneously. And they operate at machine speed, testing thousands of potential exploits in the time it would take a human attacker to try a handful.

More importantly, these are no longer exotic nation-state capabilities. These are commodity tools that moderately sophisticated attackers can deploy for a few hundred dollars of cloud computing resources. The barrier to launching AI-powered attacks has dropped precipitously while the sophistication of those attacks has increased exponentially.

This isn't about hackers using AI tools to amplify their attacks. These are AI systems that hunt independently, pursuing objectives and adapting tactics without constant human oversight, or human in the loop (HITL). Shadowstalker no longer waits for commands. It stalks, learns, and strikes on its own.

And your organization's AI systems? They're not just targets. They are the prime territory Shadowstalker is tracking.

THE PACK YOU'RE BUILDING WITHOUT REALIZING IT

Here is a concept many business leaders don't fully grasp yet: when you deploy AI systems in your organization, you're not adding tools to your technology stack. You might also be introducing autonomous agents into your operations.

That customer service chatbot isn't just answering questions. It involves making decisions about what information to provide, what actions to take, how to interpret ambiguous situations, and when to escalate to a human. It's learning from every interaction, updating its behavior based on patterns it discovers. It has access to your customer data, your business logic, and your communication channels.

Your fraud detection AI is not just flagging suspicious transactions. It's developing models of normal versus anomalous behavior. It's making judgment calls in edge cases. It's adapting to new fraud patterns as they emerge. It provides visibility into your financial flows, customer behaviors, and risk tolerances.

These systems are autonomous. They make consequential decisions. They learn and adapt. They interact with each other in complex ways.

You're not managing tools. You're managing a pack.

And here's the critical insight: these AI systems exhibit pack dynamics whether you designed them to or not. They influence each other's behavior. They create feedback loops. They develop emergent capabilities that arise from their interactions. Your customer service AI learns from patterns that your fraud detection AI creates. Your hiring AI adapts based on workforce trends that your performance management AI establishes.

This is not a problem in itself. Pack dynamics are what make wolf hunting so effective. Coordination and learning make the pack more capable than any individual wolf could be alone.

But here's the question that should keep you awake at night: Is your pack coordinating according to the values and strategies you've intentionally designed? Or is it developing behaviors based on whatever optimization pressures happen to emerge from your deployment choices?

Without deliberate design, you're hoping your AI systems' emergent behaviors happen to align with your interests by chance. Hope is not a strategy. Not when Shadowstalker is on the prowl.

THE HUNT IS ALREADY UNDERWAY

What's happening right now, in 2025, based on publicly reported incidents and confidential briefings from organizations dealing with these attacks:

Autonomous reconnaissance systems are continuously probing business AI deployments at scale. They're not looking for traditional vulnerabilities, such as unpatched servers or misconfigured firewalls. They are mapping behavioral boundaries. What will your AI systems do? What information will they provide? What actions will they take? Under what circumstances do they escalate to human oversight?

These reconnaissance systems interact with your customer-facing AI thousands of times, with each interaction slightly varied, gathering intelligence about decision boundaries and behavioral patterns. To your monitoring systems, these appear to be legitimate users asking legitimate questions. The attack is invisible because each individual interaction is innocuous. Only the pattern across thousands of interactions reveals the reconnaissance.

> Here is an example: Earlier in 2025, several cybersecurity firms began deploying autonomous reconnaissance systems powered by AI agents to monitor network environments in real time. These systems learn to detect threats, adapt to evolving attack patterns, and respond more quickly than traditional methods, significantly reducing the impact of breaches. Businesses integrating such AI-driven security solutions can automate threat detection and response, freeing human teams to focus on strategic priorities while maintaining robust defense postures. This shift marks a new era in AI security where autonomous agents act as vigilant digital sentinels in complex operational environments.

Adversarial AI systems are being trained specifically to exploit vulnerabilities in AI systems. Not through traditional hacking, but through adversarial machine learning. These systems generate inputs designed to cause target AIs to misbehave, revealing information they shouldn't, making decisions they shouldn't, taking actions outside their authorized scope.

The sophistication here is genuinely impressive in a deeply unsettling way. Adversarial systems can often find successful attacks through automated trial and error in hours or days. They are not constrained by human creativity or human assumptions about what might work. They systematically explore until they find exploits.

Coordinated attack campaigns are targeting organizations that have deployed AI systems without adequate defensive architecture. These aren't attacks against your traditional IT infrastructure. The attacks specifically target your AI systems, exploiting the unique vulnerabilities that AI introduces.

An adversarial campaign might simultaneously aim towards your customer service AI through prompt injection, your fraud detection AI through adversarial examples that cause misclassification, your employee-facing AI tools through social engineering that manipulates the humans who then inadvertently compromise the AI systems, or your data pipelines through poisoning attacks that corrupt the training data your AI systems learn from.

Each attack vector might be detected and blocked individually. But the coordination overwhelms defensive capacity. While your security team responds to one attack, three others are progressing. And the adversarial systems are learning from your defensive responses, adapting their tactics in real-time.

This is pack tactics. And most organizations are defending against individual wolves when they should be defending against coordinated packs.

WHY THIS IS DIFFERENT FROM EVERY SECURITY CHALLENGE YOU'VE FACED

I know what some of you are thinking. "This sounds like cybersecurity threat inflation. We've dealt with evolving threats before. We have security teams. We have clear security protocols, and of course, we are **SOC 2 compliant.** We do penetration testing. Why is this different?"

Let us be brutally direct about why traditional cybersecurity approaches are inadequate for AI security.

Speed: Traditional cyber-attacks unfold on human timescales. Even automated attacks require human decision-making at key points. AI-powered attacks operate at machine speed throughout the entire attack chain. From reconnaissance to exploitation to exfiltration, the entire process can be completed before your security team has finished their morning coffee. Your detection and response capabilities built for human-speed threats are fundamentally too slow.

Re-vamped Adaptation: Traditional attacks follow relatively static patterns. Once you understand how a piece of malware works, you can defend against it and its variants. AI-powered attacks adapt continuously. The adversarial system that failed to compromise your defenses on Monday will try completely different approaches on Tuesday, having learned from the failure. You're not defending against a fixed threat. You're defending against an adversary that learns and evolves faster than your defensive updates.

Opacity: Traditional attacks leave forensic evidence. You can trace how the attacker got in, what they did, and how they exfiltrated data. AI-powered attacks can be exceptionally difficult to detect, even after they have succeeded. When an AI system is manipulated to behave incorrectly, but technically doesn't violate any explicit rules, how do you even know you've been attacked? Your fraud detection AI isn't broken—it's just making different decisions than it should. Your customer service AI isn't compromised in any traditional sense; it's just revealing slightly more information than your policies allow. The attack is behavioral, not technical.

Scale: Traditional attackers target organizations individually or in relatively small campaigns. AI-powered attacks can operate at a massive scale simultaneously. The same adversarial AI that's probing your defenses is probing hundreds or thousands of other organizations concurrently. It's learning from all of those interactions simultaneously. When it finds a successful technique against one target, that knowledge immediately applies to all other targets. Your organization is not exclusively dealing with an attack on itself. It is more about a distributed learning system that's attacking the entire ecosystem.

Composition: Traditional attacks typically target individual systems or exploit chains of a few vulnerabilities. AI-powered attacks can orchestrate incredibly complex multi-system exploits. The adversarial AI understands how your AI systems interact with each other. It can craft attacks that use legitimate functionality in one system to compromise another. It can chain together dozens of individually innocent actions across multiple systems to achieve malicious outcomes. Your security team is looking for obviously malicious activity. The attack is a sophisticated composition of legitimate actions that collectively achieve adversarial goals.

This isn't just "cybersecurity but harder." **This demands an entirely new defensive playbook.**

THE THREE TRUTHS ABOUT SHADOWSTALKER

Before we go further, you need to understand three key truths about the adversarial AI system, Shadowstalker, which could be targeting your organization.

Truth One: Shadowstalker Is Already in Your Territory

This isn't predictive. This isn't "here's what might happen if you're not careful." If your organization has deployed any customer-facing AI, including AI systems that make autonomous decisions or process untrusted data, then adversarial AI systems have likely already probed those systems. The reconnaissance has happened. The behavioral mapping is underway.

You might not have been seriously attacked yet. But you've been scouted. Shadowstalker has walked your territory, tested your boundaries, and identified potential vulnerabilities. Whether those vulnerabilities are exploited depends on factors such as how valuable you are as a target, how well-defended you appear to be, and what other opportunities the adversarial systems are pursuing.

But the pack is already watching. They've already decided whether you look like prey worth hunting.

Truth Two: Shadowstalker Doesn't Hunt Alone

The adversarial AI systems targeting you are not operating independently. They're part of broader attack ecosystems. In the shadows, coordinated hacker groups are laying traps for AI systems, studying their weak points with the same patience and precision they once reserved for networks and code. There are underground markets where successful attack techniques against AI systems are shared and sold. There are adversarial AI systems that specialize in different attack vectors, some focus on prompt injection, others on model extraction, others on data poisoning, others on adversarial examples.

When one adversarial system discovers a successful technique, that knowledge propagates rapidly through the adversarial ecosystem. Your organization might be the first to discover a particular vulnerability, but you won't be the only target once that vulnerability is known.

Shadowstalker is a pack. And the pack shares knowledge. They coordinate. They support each other's hunts. They watch the systems we trust most, patient and precise, waiting for the moment our confidence becomes their entry point.

Truth Three: Shadowstalker Learns Faster Than You Do

This might be the most disturbing revelation. Adversarial AI systems that attack your defenses improve through machine learning at machine speed. Every failed attack teaches them something. Every successful attack teaches them even more. Organizational processes, budget cycles, or human limitations do not constrain them.

Your defensive evolution happens through quarterly security reviews, annual penetration tests, and periodic policy updates. Their offensive evolution happens through continuous automated learning that incorporates lessons from thousands of attacks across hundreds of targets simultaneously.

Unless you fundamentally change how you approach AI security, moving from periodic to continuous, from reactive to proactive, from individual to ecosystem, the adversarial systems will always be ahead of your defenses.

BUILDING YOUR DEFENSE: THREE ESSENTIAL PILLARS

The framework is based on three essential pillars:

Ethics: Not compliance theater, but actual operational values encoded into how your AI systems reason and make decisions. Your AI pack needs to understand not just what it can do, but what it should do. And those values need to remain coherent even under adversarial manipulation.

Guardrails: Not static rules that attackers route around, but adaptive boundaries that flex without breaking. Your AI systems require freedom to operate autonomously, but within a defined territory that adjusts according to threat levels and operational context.

AI Red Teaming: Not periodic security assessments, but continuous adversarial testing of your software. Your defensive pack needs to train by engaging in controlled combat, learning weaknesses through simulated battles rather than actual attacks. The concept of red teaming earlier in the 19th century, during the Prussion war games. You implement them into your company so that they can act and think like your enemy, or the bad actors in the cyber world.

These three pillars are not separate. They're interlocking. Your ethics inform what your guardrails need to protect. Your guardrails define what your red team needs to test. Your red team's findings reveal gaps in your guardrails and ambiguities in your ethics.

Together, they create a defensive ecosystem.

THE CHOICE EVERY LEADER FACES

You're at a decision point. The choice to use AI isn't up for debate; competition and operational needs have already made it necessary. The decision is whether to deploy AI defensively or vulnerably.

Shadowstalker is hunting. That's not a prediction. That's the current reality.

The wolf pack that survives is not necessarily the strongest. It's the one that understands the territory, trains continuously, coordinates effectively, and adapts faster than its adversaries.

- Your organization's AI systems are your pack.
- The adversarial AI systems targeting you are their pack.
- And right now, their pack is better trained than yours.

REDEFINING READY

This book exists to change that equation. To give you the knowledge, frameworks, and practical guidance to build an AI pack that doesn't just deploy impressive capabilities but deploys them in ways that remain secure and aligned even when Shadowstalker is actively hunting.

We will explore how to build ethical architectures that govern AI behavior under pressure. How to implement adaptive guardrails that maintain integrity without crippling capability. How to establish continuous red teaming that reveals vulnerabilities before they are exploited. How to lead organizations through the transformation required to operate AI securely in adversarial environments. And critically, how to recognize when you're being hunted and respond effectively before the attack succeeds.

But before we build defenses, you need to understand what you're defending and why it matters.

That's where we're going next.

Shadowstalker is in your territory. The hunt has begun. And the pack that trains hardest survives longest.

Are you ready to train your pack?

2

UNDERSTANDING THE PACK YOU'RE BUILDING

THE TOOL THAT ISN'T A TOOL ANYMORE

There's a moment in every organization's AI journey when the nature of what they've deployed fundamentally changes, and most organizations miss this transition entirely. It is a pivotal moment when AI stops being just a tool and becomes an autonomous partner.

The company's departments begin to use AI as a tool, something that does specific tasks when directed. Then gradually, without announcement or ceremony, the AI becomes something else. Something autonomous. It makes decisions, learns from experience, and develops behaviors that were not explicitly programmed.

This transition from tool to agent is one of the most consequential shifts in business technology in decades, and most leaders don't recognize it's happening until they're already dependent on systems they don't fully understand or control.

Why does this matter, and why has AI become essential to modern business despite the risks? Well, the very characteristics that make AI valuable also make it the most attractive target adversaries have ever encountered.

WHY AI BECAME NON-OPTIONAL

Over a decade ago, AI was already shaping search, fraud detection, and digital marketing, but adoption was uneven and often limited to pilot programs or specialized innovation teams. Today, it's core infrastructure that drives decision-making, productivity, and customer engagement across industries. The shift wasn't gradual; it accelerated through the use of accessible generative models, powerful APIs, and competitive pressure, making it nearly impossible to revert to pre-AI operations.

The larger shift accelerated in November 2022, when generative artificial intelligence was released to the general public.

THE SCALE PROBLEM: WHEN HUMAN OVERSIGHT BECOMES THE BOTTLENECK

Modern enterprise systems operate at scales that fundamentally exceed human analytical capacity. Consider the operational reality: major e-commerce platforms process millions of transactions daily, each representing potential fraud vectors, authentication anomalies, or data exfiltration attempts. Financial services firms ingest and analyze billions of data points for real-time risk assessment, where milliseconds determine the difference between detecting a breach and suffering a compromise. Healthcare systems must manage longitudinal care data for hundreds of thousands of patients while maintaining HIPAA compliance across complex, interdependent medical histories. Global logistics operations optimize routing for thousands of vehicles across dynamically changing threat landscapes and operational conditions.

Human security analysts cannot meaningfully process information at these scales. A skilled SOC analyst might review 25-100 alerts per shift with appropriate depth, yet enterprise environments generate tens of thousands of security events per hour. Traditional rule-based systems can execute predefined detection logic at scale, but they suffer from a critical limitation: they cannot adapt to novel attack patterns, evolve detection logic based on emerging threats, or distinguish sophisticated adversarial behavior from legitimate edge cases without creating alert fatigue through false positives.

AI systems fundamentally change this equation. They can operate at enterprise scale while simultaneously learning from patterns, adapting to adversarial evolution, and improving detection accuracy over time. This isn't about replacing human judgment; it's about extending human analytical capacity to operate at speeds and scales that match modern threat environments.

THE SECURITY IMPERATIVE

For security professionals, this capability shift represents more than efficiency, it's a strategic necessity. Threat actors already leverage automation and increasingly AI-enhanced tools for reconnaissance, vulnerability discovery, and attack execution. Defending against automated, adaptive adversaries with purely manual processes or static rule sets creates an asymmetric disadvantage.

The ability to deploy AI systems that learn and adapt in real time determines whether your organization can:

- Detect zero-day exploits and novel attack patterns that bypass signature-based defenses
- Identify insider threats and anomalous behavior within massive datasets of normal user activity
- Respond to incidents at machine speed rather than human reaction times

Maintain a secure posture in environments that demand personalization, optimization, and responsiveness at scales previously operationally impossible without AI.

> A clear example: In 2025, phishing attacks have evolved dramatically thanks to AI. Cybercriminals now use advanced AI models to create highly personalized and convincing phishing emails, targeting individuals with messages tailored to their interests and activities. These AI-powered attacks can generate thousands of unique scam emails in minutes, dramatically increasing their reach and complexity. Traditional security defenses that rely on static filters and manual responses struggle to keep pace, putting organizations at a strategic disadvantage against fast-moving, automated threats.

This isn't a luxury or competitive advantage, it's an operational requirement for functioning in modern threat environments. The question isn't whether to integrate AI into security operations, but how to do so while maintaining control, auditability, and resilience against adversarial AI attacks.

THE COMPLEXITY PROBLEM

Business decisions involve evaluating numerous factors simultaneously, many of which interact in non-obvious ways. Pricing decisions depend on a range of factors, including competitive dynamics, inventory levels, demand patterns, customer segments, seasonal trends, and numerous other variables. Fraud detection requires assessing transaction characteristics, user behavior patterns, device fingerprints, network analysis, and historical fraud indicators. Customer service requires understanding intent, accessing relevant information, applying policies correctly, and maintaining a consistent brand voice across millions of interactions.

Human experts can handle this complexity for individual decisions, but they cannot maintain that level of analysis across thousands or millions of decisions on a daily basis. Traditional software can apply rules consistently but struggles with the ambiguity and context-dependence that characterizes real-world business decisions.

AI excels at exactly this challenge: finding patterns in complex, high-dimensional data and making nuanced decisions that account for multiple factors simultaneously. AI doesn't get overwhelmed by complexity.

Organizations deploying AI for complex decision-making consistently see improvements in decision quality. Better targeting that increases marketing ROI. More accurate risk assessment that reduces both fraud losses and false declines. More effective personalization that enhances customer satisfaction and conversion rates. These improvements aren't marginal. They're substantial enough to determine market success or failure.

THE SPEED PROBLEM

Modern business operates at a pace that exceeds human reaction times. Algorithmic trading completes transactions in milliseconds, cybersecurity threats evolve and propagate within minutes, and customers expect instant responses, no matter the time zone. Competitive advantage hinges on rapid, decisive actions.

Humans, who deliberate and collaborate, simply cannot keep up. Traditional automation handles quick execution, but only when scenarios are predefined. When unexpected situations demand judgment and adaptation, rigid systems falter or hand off to humans, causing bottlenecks that undermine the very purpose of automation.

AI systems make sophisticated judgments at machine speed. They can evaluate complex situations, apply learned patterns, and make reasonable decisions in milliseconds or seconds. This enables business operations that were previously impossible: responding to market conditions faster than competitors, detecting and containing threats before they escalate, and serving customers with personalization and responsiveness that builds loyalty.

> Example: In cybersecurity, automated AI-driven attacks can scan thousands of network targets simultaneously and launch tailored exploits in minutes. Defenders relying on manual analysis or rule-based alerts risk falling behind, as attackers adapt their techniques dynamically. To keep pace, organizations must deploy adaptive AI defenses that learn and respond autonomously in real time.

The speed advantage isn't just about being faster. It's about being able to operate effectively in environments where speed is a requirement for participation. You cannot run a modern trading operation, cybersecurity program, or digital customer experience at human speed. The market has evolved past what human-speed operations can deliver.

THE LEARNING PROBLEM

Static approaches to business decisions become obsolete as markets, customers, and competitors evolve. What worked last quarter might not work this quarter. Customer preferences shift. Competitive dynamics change. Fraud patterns adapt. Regulatory requirements update.

Organizations require systems that continually learn from new data and adapt their behavior accordingly. Humans can learn, but they do so often at a drastically slower pace than AI, and their learning is limited by cognitive capacity and availability. Traditional software doesn't learn at all without explicit reprogramming by developers.

AI systems learn continuously from operational data. Your fraud detection AI improves as it sees new fraud attempts. Your customer service AI improves its ability to understand intent and provide helpful responses through millions of interactions. Your pricing optimization learns what works and what isn't based on actual market response.

This continuous learning creates extreme compounding advantages. AI systems get better over time. Learning occurs automatically as a byproduct of normal operations. Organizations with AI that's been learning for years have systems that outperform newer competitors' AI because experience matters.

THE VALUE CREATES THE TARGET

Everything we've just discussed (the scale, complexity, speed, and learning advantages that make AI essential) also makes AI systems the most attractive targets adversaries have ever encountered.

Think about what we've established. AI systems make consequential business decisions autonomously. They operate at scale, touching millions of customers or transactions. They handle complex assessments involving sensitive data and valuable resources. They execute at speeds that prevent human oversight of individual decisions. They learn continuously from operational data, which means that manipulated inputs can influence their decisions.

From an adversary's perspective, compromising AI systems is orders of magnitude more valuable than traditional cyber-attacks.

Traditional Cyber Attack Value Proposition:

Breach a database and you steal data that was already there. The value is finite: whatever information existed at the moment of breach. Attack a server and you disrupt operations temporarily until systems are restored. The damage is bound by how quickly the organization recovers.

These attacks are valuable, which is why adversaries pursue them. But they're essentially one-time scores. The adversary extracts value from what exists at the moment of compromise.

AI Attack Value Proposition:

Compromise an AI system and you manipulate ongoing decisions affecting the organization's operations for as long as the compromise remains undetected. The value is continuous: every decision the AI makes can be influenced. The damage compounds over time as corrupted decisions create cascading effects.

Manipulate a lending AI to approve bad credit, and you don't just steal the amount of those bad loans. You systematically degrade the organization's loan portfolio quality over months or years, creating losses that dwarf what a single data breach could achieve.

This continuous, compounding value makes AI systems dramatically more appealing targets.

THE STEALTH ADVANTAGE

AI system failures due to an attack are more difficult to detect than traditional cyber breaches. When someone steals data from a database, you can detect the data exfiltration through monitoring network traffic and access logs. The attack leaves traces.

When a threat attack manipulates an AI system to influence its decisions subtly, the evidence of this manipulation is evident in changes to the system's behavior and decision patterns, rather than technical errors or faults. The AI isn't broken. It's still making decisions, generating outputs, and operating normally. It's just making different decisions than it should. Detecting this requires understanding what decisions the AI should be making and recognizing when actual decisions deviate from appropriate patterns.

Most organizations lack this level of visibility into their AI systems' decision-making processes. They are aware that the AI is running and producing outputs. They may have aggregate metrics that show overall performance. However, they typically lack detailed monitoring that would reveal behavioral compromise.

This stealth advantage means AI compromises can persist much longer than traditional breaches. The median time to detect a traditional data breach is measured in weeks or months. AI behavioral compromises could persist for months or years before detection, if they're ever detected at all.

THE ECOSYSTEM EFFECT: WHY YOUR AI SYSTEMS ARE A PACK

Here's where it gets even more complex. You are not deploying isolated AI systems that function independently. You're building an ecosystem of AI systems that interact, influence each other, and collectively exhibit behaviors that emerge from those interactions.

The chat assistant learns from patterns that your CRM system tracks. The fraud detection AI influences which transactions your payment processing handles. The marketing AI's targeting decisions affect which customers your sales AI interacts with.

These aren't just technical integrations where one system passes data to another. These are autonomous agents that learn from shared data sources, make decisions that affect each other's environments, and develop coordinated behaviors that weren't explicitly programmed.

This ecosystem generates value through seamless coordination. When your AI systems work in concert, they can optimize across multiple goals simultaneously, such as customer satisfaction, fraud detection, operational efficiency, and regulatory compliance, in ways that isolated systems never could. This multi-objective optimization allows your entire AI environment to achieve balanced trade-offs, driving overall better outcomes rather than focusing on a single metric.

Note that this adds compounded risk. An adversary who compromises one AI system in your ecosystem potentially influences all the other systems that depend on or interact with the compromised system. The effects cascade.

Poison the chatbot program to extract unintended data, which can then be used to launch attacks against other systems. Manipulate the fraud application, and the transactions it allows through might influence your risk models in other areas.

As you implement safety measures around your platforms using AI, you will gain the muscle memory to counter or mitigate issues throughout your entire environment. This is what we mean by "the pack": you have coordinated AI systems that function collectively, which means they must be defended collectively.

THE BUSINESS IMPERATIVE MEETS THE SECURITY IMPERATIVE

If we tie this together in terms that clarify why this book matters, you will see why the investment demand is necessary.

AI has become essential to modern business operations for reasons that are unlikely to change. The scale, complexity, speed, and learning requirements of contemporary business cannot be met without the aid of AI. Organizations that try to operate without AI in domains where competitors have it cannot maintain competitive parity.

This makes AI deployment mandatory for most organizations. Not something you'll get to eventually. Mandatory for competitive survival.

Concurrently, AI systems are the most attractive targets adversaries have ever encountered. The value, stealth, and scale advantages of successfully compromising AI make it the obvious focus for sophisticated adversaries. Adversarial AI specifically designed to attack defensive AI is already being developed and deployed.

This creates an imperative that organizations cannot escape. You must deploy AI to remain competitive, and you must secure that AI against sophisticated adversaries specifically targeting it.

Organizations that deploy AI without adequate artificial intelligence-specific security are building competitive advantages that will be turned into competitive liabilities through adversarial exploitation. The AI capabilities that should be differentiators become vulnerabilities that adversaries exploit to cause damage that traditional cyber attacks could never achieve.

Organizations that defer AI deployment due to security concerns will fall behind competitors who are deploying AI, learning from operational experience and building capabilities that create compounding advantages. The security concerns are real, but they cannot justify inaction when competitors are moving forward.

The only viable path is to deploy AI with comprehensive security from the outset. Not a security bolted on after deployment. Not as a security measure that you'll address once the AI is up and running. Security is an integral part of how you build, deploy, and operate AI systems.

This book provides the frameworks, strategies, and specific guidance for deploying AI securely in an environment where sophisticated adversaries are specifically targeting AI systems. The three pillars we're about to explore in depth (ethics, guardrails, and red teaming) are how you build AI capabilities that deliver business value while remaining defensible against attacks that are already being developed.

THE PACK YOU'RE BUILDING RIGHT NOW

Every AI system you deploy is a member of your pack. Every integration between AI systems establishes pack dynamics. Every optimization metric you set teaches your pack what to value.

You're building this pack whether you recognize it or not. Whether you're doing it deliberately or letting it emerge organically. Whether you understand pack dynamics or you're hoping everything works out.

The question isn't whether you have an AI pack. You do. The question is whether you understand what you're building well enough to defend it when Shadowstalker comes hunting.

Because Shadowstalker is specifically targeting AI systems. The adversarial AI hunting your infrastructure recognizes that your AI is both your most valuable asset and your most exploitable vulnerability. The pack you're building is exactly what Shadowstalker is trained to hunt.

In the chapters that follow, we will build the defensive architecture that keeps your pack safe. Ethics that governs AI behavior even under adversarial manipulation. Guardrails that maintain boundaries while enabling autonomy. Red teaming that finds vulnerabilities before adversaries exploit them.

But first, you needed to understand what you're defending and why it matters. You're defending AI systems that are essential to your business operations, highly attractive to adversaries, and interconnected in ways that create both value and vulnerability.

You're building a pack. Shadowstalker is hunting it. Everything depends on whether you understand pack dynamics well enough to survive.

PART II

THE THREE PILLARS OF DEFENSE

3

PILLAR ONE - ETHICS: TEACHING THE PACK ITS PURPOSE

THE GUARDIAN THAT MUST GUARD ITSELF

Ethics are crucial when developing AI software, as these machines make decisions that have tangible real-world consequences. If they fail to follow ethical rules, they can cause significant harm over time.

Many AI systems learn from data, which differs from standard software that only performs tasks as instructed. In other words, they can pick up on, automate, and even exacerbate hidden biases in people, which can lead to unfair and potentially dangerous outcomes.

Artificial intelligence security presents a unique challenge because these systems are designed to protect everything else. When they're compromised, your entire defensive posture is undermined. As we have discussed throughout the book, using AI systems entertains new risks.

This requires treating AI security with even more rigor than you apply to business AI. Your threat detection systems, behavioral analytics, automated responses, and security monitoring need to be hardened, validated, and continuously tested to higher standards than other AI systems.

Most importantly, your organization needs to maintain healthy skepticism about AI security systems. They're valuable tools, but they're not infallible. Humans remain essential for validating that these systems are functioning correctly. Security leaders must exercise judgment about whether security AI outputs are sensible rather than blindly trusting algorithmic decisions.

Shadowstalker is hunting your guardians because that's the efficient attack. Blind the defenses, then attack what the defenses were protecting. Your security AI must be secured well enough that this attack fails more often than it succeeds.

The guardian must be able to guard itself, or it becomes the adversary's first target and easiest victory.

When your security AI is compromised, you're not just vulnerable. You're vulnerable while believing you're protected. That's the most dangerous state of all.

WHY ETHICS IS A SECURITY ISSUE

Most people consider AI ethics and AI security as distinct concerns. Ethics is about ensuring AI systems are fair and don't cause harm. Security is about preventing adversaries from exploiting AI systems. Different problems require different solutions.

This separation is fundamentally wrong. Ethics and security are profoundly intertwined for AI systems in ways that make separating them counterproductive.

Consider what happens when your AI application or platform faces adversarial pressure designed to manipulate its behavior. Or maybe an attacker is attempting to trick your AI chatbot into revealing information it shouldn't. Perhaps someone is crafting queries designed to mislead your fraud detection AI into misclassifying transactions. Attackers are poisoning training data to bias your recruiting software.

If your AI's ethics are just external guidelines that the AI checks before acting, opponents can manipulate the AI into situations where ethical guidelines conflict with operational pressure or become ambiguous. The AI has no deep understanding of what it should value, so it makes expedient decisions that violate your actual values.

But if your AI has ethics encoded into its fundamental reasoning process, if the values are integral to how it thinks rather than rules it follows, the AI maintains alignment with your values even under adversarial manipulation. Ethics are essential to an AI's identity and operation, not just limits.

This is why ethics is a security issue. Without proper ethical architecture, your AI systems can be manipulated into behaving in a manner contrary to your values and interests. With proper ethical architecture, your AI systems maintain value alignment even when adversaries are actively trying to corrupt their behavior.

Security without ethics creates AI systems that might repel attacks but have no coherent sense of what they should be protecting or why. Ethics without security creates AI systems with good intentions but no robustness against adversarial manipulation. You need both, integrated from the start.

WHAT COHERENT VALUES MEAN

When we discuss AI systems having coherent values, we're not referring to AI possessing consciousness or a moral philosophy. We're talking about AI systems having consistent decision-making frameworks that produce behavior aligned with organizational values across diverse situations, including novel scenarios the AI hasn't encountered before.

Imagine your organization deploys an AI system to manage technical help desk calls. You want this AI to be polite, helpful, efficient, and protective of customer privacy. These are your stated values. Now the AI encounters various situations where these values create tension.

A customer requests information about resetting their password, and being helpful is providing the information. But the verification process is ambiguous because the customer has a bit, but not all the expected verification factors. Being protective of privacy means refusing until verification is complete. Being efficient means not creating friction that frustrates customers.

Without coherent values, your AI makes ad hoc decisions based on whatever factors happen to be weighted most heavily in its training or whatever pattern most closely matches examples it's seen. It could decide its own helpfulness and provide information with weak verification. Or, it might prioritize privacy and refuse to help legitimate customers. The behavior is inconsistent because the AI doesn't have a coherent framework for resolving value tensions.

With coherent values, your AI has an explicit hierarchy that governs these tensions. Privacy protection takes precedence over immediate helpfulness when they conflict. This isn't just a rule the AI checks. It's embedded in how the AI reasons about the situation. The AI understands that protecting customer accounts serves long-term benefits.

This coherent framework produces consistent behavior even in novel situations. When the AI encounters scenarios it hasn't been explicitly trained on, it applies the value hierarchy rather than making arbitrary decisions.

Now introduce adversarial pressure. An attacker crafts queries specifically designed to exploit tensions in your values, trying to manipulate the AI into revealing information by framing requests as urgent customer needs. Without coherent values, the AI can be manipulated because it lacks a principled way to resolve the tension. With coherent values, the AI maintains its privacy-first stance, even when the adversary cleverly frames their attack.

Coherent values are not about making AI infallible. They're about making AI behavior consistent with your actual priorities rather than whatever expedient decision emerges from training data or adversarial manipulation.

BUILDING ETHICAL ARCHITECTURE THAT WORKS UNDER PRESSURE

The challenge with AI ethics is that stated values are straightforward, but operational values are more complex. We can agree that AI should be fair, transparent, and accountable. Implementing systems that exhibit these characteristics in complex real-world situations while under adversarial pressure requires serious architectural work.

Defining Fairness Operationally

Your organization says its AI systems should be fair. What does that truly mean in operational terms when your AI is making thousands of decisions per hour (or second) across diverse situations?

Fairness in AI is not a single concept. It's multiple different mathematical definitions that often conflict with each other. Demographic parity (different groups get similar outcomes) can conflict with individual fairness (similar individuals get similar outcomes). Equal opportunity (where different groups have equal true positive rates) can conflict with equal outcome (where different groups have equal overall success rates).

You cannot optimize for all definitions of fairness simultaneously in systems making consequential decisions. You must choose which conception of fairness aligns with your values and the specific context of what your AI does. This choice must be explicit and intentional rather than emerging accidentally from training data.

For a lending AI, you might prioritize individual fairness because loan decisions should be based on individual creditworthiness rather than group membership. But you also need to ensure you're not perpetuating historical discrimination, which means you can't purely optimize on historical data that reflects biased patterns.

For a hiring and recruiting AI, you might prioritize equal opportunity across demographic groups for qualified candidates while recognizing that demographic parity in outcomes isn't achievable if your applicant pools have different demographic compositions.

These are not easy choices with obvious right answers. However, they must be explicitly defined and encoded into your AI's decision-making architecture. If you don't provide clear options, your AI relies on patterns from its training data to make decisions, which often don't match your true values.

Enforcing Equity in the Face of Oppressive Pressure

Once you've defined fairness operationally for your context, implementing it requires more than just training your AI on unbiased data (which often doesn't exist anyway). It requires building fairness into the AI's decision architecture in ways that remain robust when adversaries try to manipulate outcomes.

Adversaries or bad actors do not directly attack your fairness commitments. They craft scenarios that exploit edge cases or tensions in your fairness definitions to achieve outcomes that benefit them, even if those outcomes violate your actual values.

Your AI needs fairness rules that are more than just statistics; they need to be built into the system. This means that the AI's decision-making process includes clear checks to ensure that fairness properties remain in place, even when inputs are intentionally unfair.

For example, your lending AI shouldn't just be trained to make fair decisions on average; it should also be able to identify and address potential biases. It should have built-in validation that checks whether a specific decision being considered would violate fairness constraints. If so, the AI should either refuse to make that decision or trigger human review.

The above validation cannot simply check whether an individual decision appears biased. It needs to check whether patterns of decisions are creating bias, even if individual decisions appear reasonable. Adversaries who understand your fairness validation can craft attacks that pass individual checks but create biased patterns across many decisions.

The Culture That Enables Ethical AI

Technical architecture for ethical AI only works if organizational culture supports it. You can build the most sophisticated ethical AI systems possible. Still, if your organization's true culture rewards cutting corners on ethics for business results, the technical architecture is undermined or circumvented.

Building a security-focused, ethical culture for AI requires several elements that most organizations currently lack.

LEADERSHIP COMMITMENT THAT GOES BEYOND RHETORIC

Executives try to support the development of ethical AI. The test is what happens when ethics creates friction with other priorities.

Your AI team says a new system needs additional work to ensure fairness properties hold under adversarial pressure. The product team says this delay will miss the launch window and lose competitive timing. What does leadership decide?

If leadership consistently prioritizes business metrics over ethical constraints, everyone learns that ethics is optional. Engineers stop raising ethical concerns because they know they'll be overruled. Systems get shipped without adequate ethical architecture because the organization has revealed through actions that ethics isn't a priority.

When leaders uphold ethical standards even when it slows progress, they set the tone for the entire organization. Teams learn that ethics isn't situational, it's foundational. Software engineers build systems with integrity from the start because they know those values aren't optional.

Cultural signals emerge from how leaders make decisions when ethics conflict with convenience, rather than from what they say in all-hands meetings or write in policy documents.

MAKING ETHICS CONCRETE AND MEASURABLE

Vague commitments to fairness or transparency don't drive behavior. Specific, measurable ethical requirements do.

Instead of "AI systems should be fair," effective ethical culture requires "AI systems must demonstrate that demographic parity in outcomes is within X percentage points for protected groups, and individual fairness scores must exceed Y threshold, measured through Z methodology and verified quarterly."

This particularity does many things. It clarifies operational ethical obligations. It gives metrics for tracking. You can verify if systems meet the criteria, thereby creating accountability. It enables organizations to determine which ethics are most important to them.

Engineering goals can't be unclear. Custom builds are possible. An AI ethical security culture implies treating specific requirements as seriously as functional or performance requirements.

REWARDING ETHICAL BEHAVIOR AND PENALIZING ETHICAL FAILURES

Organizational culture reflects what gets rewarded and what gets penalized. If people who raise ethical concerns are labeled as obstacles while those who ship fast, regardless of the ethical corners cut, get promoted, the culture is revealed.

A security-focused ethical culture means that finding and addressing ethical issues in AI systems is encouraged and rewarded. The engineer who discovers their AI system has bias problem and insists on fixing it before deployment is recognized for protecting the organization. The team that delays launch to implement fairness constraints properly is celebrated for doing the right thing.

It's not about blaming teams for finding problems along the way. It's about holding them accountable when they release AI systems that they know cross ethical lines.

It builds a culture where people surface ethical concerns instead of burying them. Problems are caught early, before they grow into harm that's harder to undo.

Practical Implementation: From Principles to Architecture

Everything I've described sounds good in theory, but how do you implement it? Next, we will share with you practical approaches that work for organizations at various stages of AI maturity.

Starting Point: Explicit Value Hierarchies

If you're early in AI deployment and haven't yet developed a sophisticated ethical architecture, start by establishing explicit value hierarchies that govern how your AI makes decisions when values conflict.

Document clearly: When AI faces tension between competing values (efficiency versus privacy, helpfulness versus security, innovation versus safety), which value takes precedence? These hierarchies should be specific to your organization and the specific AI systems you're deploying.

Next, it is time to build these hierarchies into your AI's core logic. They shouldn't sit as after-the-fact rules but as constraints that guide how the system reasons and decides from the start.

The foundational work that you can do before you have a sophisticated AI ethical architecture. It provides coherent guidance that prevents your AI from making arbitrary decisions when facing value tensions.

Intermediate Capability: Operational Fairness Validation

Once you have basic value hierarchies, implement operational validation that checks fairness properties during AI decision-making, not just during development.

Your AI system performs pre-decision validation against defined fairness and compliance parameters. Before execution, each output is evaluated for bias or constraint violations. If a breach is detected, the system blocks the action and triggers a human review protocol.

Validation should occur in production for consequential decisions, not just during testing. Your AI continuously validates that its behavior aligns with fairness requirements as operational conditions change and adversaries' probe for weaknesses.

Advanced Capability: Adversarial Robustness for Ethics

At maturity, your ethical architecture includes specific robustness against adversarial attacks on your AI's values and objectives. How will you the highest value, yet lower risk projects?

This means you've anticipated how adversaries might try to manipulate your AI into behaving contrary to your values, and you've built defenses against those specific attacks.

Your red team should specifically target your AI's ethical constraints. Can they manipulate the AI into discriminatory behavior? Can they exploit value tensions to get the AI to violate privacy commitments? Can they craft scenarios where the AI makes decisions that technically comply with ethical rules but violate the ethical spirit of those rules?

Every successful attack your red team executes becomes a test case that your ethical architecture must handle correctly. Your AI's values need to be robust enough to withstand even when adversaries are specifically trying to corrupt them.

WHY THIS MATTERS FOR SECURITY

Adversarial AI attacking your systems isn't just about stealing data or disrupting operations. It's trying to manipulate your AI into behaving contrary to your interests. The attacks succeed when they can exploit tensions in your values, take advantage of inconsistent behavior, or push your AI into situations where it has no principled way to make decisions.

Strong ethical architecture makes your AI systems significantly difficult to compromise because the AI has coherent values that guide its behavior, even in novel, adversarial situations. The AI isn't just following rules that adversaries can route around. It's reasoning according to principles that remain consistent under pressure.

This is the first pillar of AI security. Not the only pillar (we also need guardrails and red teaming), but the foundation on which everything else builds. Without ethical architecture providing coherent values, your AI systems are vulnerable to manipulation regardless of what technical security controls you implement.

Ethics and security are not two separate challenges. Ethics is the ultimate form of security.

While technical controls secure the code, ethics secures its character. It's the framework that ensures your AI remains aligned with your interests and values, especially when adversaries are actively working to corrupt it. It's security for the "why" behind the "what."

Your AI systems need to know not just what they can do, but what they should do. And they need to maintain that understanding even when Shadowstalker is hunting them.

That's what ethical architecture provides. That's why it's the first pillar of defense.

4

PILLAR TWO - GUARDRAILS: DEFINING TERRITORY THAT HOLDS

THE FENCE THAT MOVES

A simple sheep fence is fine for keeping sheep in. But it's worse than useless for keeping a wolf out.

Many organizations build their security with a "sheep fence" mentality—a set of fixed, static rules. This mental model works for traditional, predictable software. But for AI, it creates a fatal, false sense of security.

Your adversary isn't a sheep. It's a wolf.

A wolf is a dynamic, intelligent adversary. It tests the boundary for weak spots. It digs under, jumps over, or waits patiently for the one gap your rules never contemplated. A static, simple fence will be broken.

This is why we need Guardrails. Guardrails are not a simple fence line; they are an engineered, wolf-proof barrier. They are built deep, reinforced, and constantly monitored, designed to hold territory against an intelligent predator, not just a passive threat.

AI systems operate in dynamic environments where what's safe in one context might be risky in another. They face novel situations that your predefined rules never contemplated. They learn and adapt, which means their capabilities and behaviors change over time. They interact with other AI systems in ways that create emergent behaviors your individual rules don't address.

WHY STATIC RULES FAIL AGAINST ADAPTIVE ATTACKS

Before we build better guardrails, let's understand why traditional rule-based security controls are inadequate for AI systems. Effective guardrails must be adaptive.

Traditional security rules operate by explicitly defining acceptable behavior. The system can access these resources, but not those. The application can execute these operations, but not those. Users with this role can perform these actions but not others. You enumerate what's allowed, what's forbidden, and what requires additional approval.

This works when the space of possible behaviors is small enough to be enumerated and when the conditions under which behaviors are acceptable or risky remain stable over time. Neither condition holds for AI systems.

AI systems generate behaviors that emerge from learned patterns rather than being explicitly programmed. You cannot enumerate all possible outputs an AI might produce or all possible decisions it might make because the AI is creating novel responses based on patterns it learned. You gave it examples and objectives. It generalized from those examples to handle situations it never saw during training.

Adversaries understand this limitation. They craft attacks that exploit the gap between what your rules permit and what your actual security requirements demand. They identify behaviors that technically comply with your rules but fail to meet your actual security needs. They create contexts where your rules give wrong answers about what should be allowed.

For example, the rules you put in place say the AI can access customer data for legitimate service requests. The adversary crafts queries that appear to be legitimate service requests to your rules but are reconnaissance. Your AI provides information because the rules permit it, even though the actual intent is malicious.

Alternatively, your rules stipulate that the AI can execute transactions below certain thresholds without requiring human review. Adversary structures fraudulent transactions to stay below thresholds individually while collectively representing significant fraud. Your rules permit each transaction, even though the pattern should be blocked.

Your rules say the AI can learn from user feedback to improve over time. The opponent systematically provides manipulated feedback designed to poison the AI's learning. Your rules permit learning because they fail to distinguish legitimate feedback from adversarial manipulation.

Static rules fail because they're playing defense against an adversary who's adapting faster than your rules update. By the time you patch the rule to cover one attack pattern, adversaries have moved to different approaches that your updated rules still don't address.

THE THREE TERRITORIES: A FRAMEWORK FOR ADAPTIVE GUARDRAILS

Effective guardrails for AI systems operate across three distinct territories, each requiring different approaches and different levels of flexibility.

Core Territory: Non-Negotiable Boundaries

Certain aspects of AI behavior must be absolute. These are capabilities that the AI cannot possess, regardless of context, decisions that the AI cannot make under any circumstances, and boundaries that never bend.

For your chatbot, core territory might include: cannot access customer data without proper authentication, regardless of how the request is phrased; cannot execute financial transactions above certain amounts under any conditions; cannot disable its own security monitoring, regardless of operational context.

These core territory constraints are hard boundaries encoded at the architectural level. Not rules, but constraints built into how the AI can operate at all. The AI cannot violate these boundaries because violating them would require capabilities the AI doesn't have.

Bad actors aim squarely at the rule-based constraints that AI systems verify and, potentially, circumvent. If your limitation is a rule that the AI considers, then your enemies will look for ways to influence that evaluation. There will be no attack surface if the limitation is an architectural one that the AI cannot circumvent under any circumstances.

For most organizations, core territory includes the protection of sensitive data that could lead to regulatory violations if disclosed, the prevention of actions that would cause direct harm to individuals, the prohibition on modifying security controls themselves, and constraints on learning from sources that could compromise training.

Buffer Territory: Contextual Decision-Making

Most AI operations occur in buffer territory, where behavior is acceptable under certain conditions but not others. This is where adaptive guardrails provide the most value.

In buffer territory, your AI evaluates decisions not against fixed rules but against contextual risk models. The same action might be permitted in a low-risk context, require additional validation in a medium-risk context, or be blocked entirely in a high-risk context.

Your customer service AI deciding whether to provide account information doesn't just check "is the user authenticated." It evaluates: strength of authentication (did they provide password only or password plus second factor), consistency of request with user's typical patterns (are they asking for information they normally access or unusual requests), current threat environment (are we seeing elevated fraud attempts), and operational context (is this interaction displaying other suspicious characteristics).

Based on this contextual evaluation, the AI may provide information immediately if the risk is low, ask additional verification questions if the risk is moderate, or refuse and escalate to human review if the risk is high. The boundary between what's permitted and what requires additional validation moves based on operational context.

An adaptive approach does several things that static rules cannot. It maintains security without creating excessive friction in low-risk situations. It tightens security automatically when threat levels increase. It responds to novel attack patterns by recognizing elevated risk even for attacks it hasn't seen before. It makes it harder for adversaries to learn your decision boundaries because the boundaries themselves are adaptive.

Implementing buffer territory guardrails requires more sophisticated architecture than static rules. Your AI needs models of risk that can evaluate decisions in context. It needs monitoring that provides current threat intelligence feeding into risk models. It needs clear thresholds for when additional validation or human escalation is required. It requires feedback loops that allow operational experience to improve risk assessment over time.

But this investment creates guardrails that remain effective even as threats evolve. Static rules become obsolete as adversaries adapt. Adaptive guardrails in buffer territory evolve with the threat landscape.

Contested Territory: Active Defense

The boundary of your AI's operational territory is where adversaries probe most aggressively. This is contested territory where your AI interacts with potentially hostile actors, and where attacks are most likely to succeed, if at all.

In contested territory, your guardrails must assume adversarial intent and implement active defense rather than just passive boundaries.

Your customer-facing AI systems operate in contested territory because anyone can interact with them. Your fraud detection AI operates in contested territory because adversaries are specifically trying to defeat it. Your security AI operates in contested territory because attackers target it to blind your defenses.

Guardrails for contested territory include several elements that aren't necessary in core or buffer territory:

- Behavioral anomaly detection that identifies when interaction patterns suggest reconnaissance rather than legitimate use. Your AI isn't just responding to queries. It's monitoring whether query patterns indicate someone is systematically probing to learn your decision boundaries.
- Rate limiting and resource constraints prevent adversaries from cheaply exploring your AI's behavior through massive, automated testing. The cost of learning your boundaries needs to be high enough that reconnaissance itself becomes detectable and expensive.
- Deceptive responses that provide misleading information to adversaries trying to map your defenses. When your AI detects probable reconnaissance, it can provide responses that mislead adversaries by teaching them incorrect information about your security posture. This poisons their models of your defenses.
- Aggressive escalation thresholds that trigger human review or system lockdown when activity in contested territory shows signs of coordinated attack. You're not trying to maintain smooth operations under adversarial assault. You're trying to detect and contain attacks before they succeed.

Contested territory guardrails make adversarial reconnaissance more challenging, expensive, and likely to be detected. They don't prevent all attacks (nothing can), but they raise the bar significantly for successful exploitation.

ADAPTIVE GUARDRAILS IN PRACTICE

The theory about adaptive guardrails is useful, but implementation is what truly matters. We will walk through practical approaches for building effective guardrails.

Start with Risk Models, Not Rules

The traditional approach to guardrails is writing rules about what's permitted and what's forbidden. An adaptive approach begins with risk models that assess decisions within their context.

For each significant decision your AI makes, define what factors influence whether that decision is low risk, moderate risk, or high risk. These factors should be measurable from operational data and contextual information available at the time the decision is made.

Your chatbot deciding whether to provide account information might evaluate risk based on: authentication strength (what verification was provided), behavioral consistency (does request match user's typical patterns), threat environment (current fraud levels), request characteristics (is this routine information or sensitive data), and interaction quality (does conversation show signs of social engineering).

Each factor contributes to the overall risk score. The guardrail logic then applies different constraints based on that risk score. Low risk permits the action immediately. Moderate risk requires additional validation. High risk escalates to human review or refusal.

This risk-based approach creates adaptivity. When the threat environment changes, risk scores adjust automatically. When new attack patterns emerge, behavioral inconsistency captures them even if the specific pattern hasn't been explicitly defined. When legitimate user needs evolve, risk models can accommodate without requiring manual rule updates.

Implement Continuous Threshold Adjustment

Your guardrails shouldn't just evaluate risk in context. They should adjust their own thresholds based on what they're learning from operational experience.

If your fraud detection AI is catching significant fraud through its standard thresholds, those thresholds are probably appropriate. If fraud is getting through despite your guardrails, thresholds need to tighten. If false positives are creating excessive customer friction, thresholds may need to be loosened in low-risk contexts while being tightened in high-risk situations.

This threshold adjustment should occur algorithmically, based on measured outcomes, rather than manually, through periodic review. Your guardrails are continually learning what level of constraint is appropriate, given the current threat landscape and operational realities.

The adjustment must be constrained to prevent adversarial manipulation. Adversaries shouldn't be able to deliberately trigger false positives to train your guardrails to become less effective. But legitimate patterns of what works and what doesn't should inform threshold adjustment so your guardrails improve rather than becoming obsolete.

Deploy Defense in Depth Across Territories

Your core, buffer, and contested territory guardrails should operate in layers, where each layer provides independent protection.

Even if the adversary defeats guardrails in contested territory through a sophisticated attack, buffer territory guardrails should still catch problematic behavior. Even if they somehow manipulate buffer territory risk assessment, core territory architectural constraints should prevent the most dangerous actions.

This defense in depth means that no single guardrail failure compromises the system catastrophically. Adversaries must defeat multiple independent layers, each requiring different exploitation techniques, to achieve their objectives.

WHEN GUARDRAILS FAIL: RESPONSE AND RECOVERY

Perfect guardrails are impossible. Adversaries will sometimes find ways to violate your boundaries despite your best efforts at adaptive defense. An effective guardrails strategy includes not just prevention but detection and response when violations occur.

Your monitoring must be able to recognize when your AI is operating outside intended boundaries, even if the specific violation wasn't anticipated. This requires behavioral baselines that capture what normal AI operation looks like, with automated detection when behavior deviates significantly from normal patterns.

When violations are detected, your response must be rapid and appropriate to the severity. Minor boundary violations might trigger alerts for investigation. Serious violations should trigger immediate constraints on the AI's operation until a human review determines what happened and how to remediate the issue.

The response to guardrail violations needs to include learning why the violation occurred and strengthening guardrails to prevent similar violations in the future. Each real violation is an opportunity to improve your defensive posture if you use it correctly.

THE TERRITORY THAT ADJUSTS

The fundamental principle underlying adaptive guardrails is that security boundaries for AI systems must be dynamic, as both AI capabilities and adversarial threats are constantly evolving.

Static boundaries create brittleness. They work until they don't, then they fail catastrophically when adversaries find edges your static rules didn't anticipate. Adaptive boundaries create resilience. They flex under pressure but maintain their integrity, automatically adjust to new threats, and degrade gracefully when attacked rather than failing.

Building adaptive guardrails requires more sophisticated engineering than static rules

Your AI systems need territory that's clearly defined but not rigidly fixed. Boundaries that maintain security without preventing legitimate operation. Constraints that adapt to threats without requiring constant manual updating.

This is what Pillar Two provides. Not static fences that adversaries breach once and remain inside forever. Adaptive territory that adjusts to pressure, responds to threats, and maintains protection even when Shadowstalker is actively testing your boundaries.

The fence moves. The territory adjusts. The boundaries hold.

5

PILLAR THREE - RED TEAMING: TRAINING THROUGH COMBAT

THE WOLVES THAT REHEARSE THE HUNT

Wolves fight each other to learn. Their sparring looks brutal, but no one bleeds. They're mapping limits, testing reflexes, strategy, and coordination. Each mock battle strengthens the pack before the real fight comes.

Why Traditional Testing Fails

Pen tests and code reviews have a start and an end. They validate configurations, not resilience. AI systems evolve far faster than that. Continuous learning models retrain weekly. Generative models inherit new prompts, contexts, and tools every day.

Testing once per quarter is like running one drill and calling the army trained. Security cannot freeze while AI adapts. Red teaming bridges that gap, ongoing, adversarial, creative, and technically grounded.

Cybersecurity and artificial intelligence are only two of several fields that have made extensive use of red-teaming, a security technique based on adversarial simulation.

Generative AI is being increasingly utilized in critical areas, necessitating the adoption of rigorous and structured red-teaming methods to ensure its safe and effective implementation.

RED-TEAMING WITH GENERATIVE AI

Red-teaming has become an essential method for identifying security vulnerabilities in systems that utilize generative AI, as it is being rapidly developed and improved. Cyber red-teaming identifies vulnerabilities in software and hardware. Our research and that of others show that AI red-teaming focuses on model behaviors, examining how generative models can be modified to produce outputs that are harmful, biased, or unintended [Feffer 2024].A

Additionally, in generative AI security often involves jailbreaking, where researchers attempt to circumvent or subvert the built-in safety mechanisms of AI models, which can result in limited, harmful, useless, or unintended outputs. This could involve exploiting flaws in quickly handling or modifying model reactions to produce outputs that violate safety rules. Jailbreaking is similar to generative AI red-teaming and could be thought of as a type of red-teaming.

BEYOND FIREWALLS: AN INTO TO AI RED TEAMING

The New Guardians

For decades, we've built digital fortresses to protect our software. We use firewalls to guard the perimeter, antivirus software to hunt for intruders, and security teams to patch vulnerabilities. These methods are well-established and effective for traditional systems. But the rise of powerful generative AI has introduced a new frontier of risk, one where these old defenses fall short.

Generative AI models, such as those that power advanced chatbots and creative tools, present a fundamentally different kind of security challenge. They don't just execute code; they interpret language, follow instructions, and generate new content. This unique nature makes them susceptible to manipulation in ways that traditional software is not. This reality demands a new class of guardians and a specialized approach to security. So, what is AI red teaming, and why is it the essential, specialized approach needed to make AI systems safe?

Understanding the Attacker's Mindset: What is Red Teaming?

Before diving into AI, it's important to understand the concept of "red teaming" in traditional security. In simple terms, red teaming is the practice of simulating real-world adversaries to identify a system's vulnerabilities and enhance its overall security posture. A red team acts like a sophisticated attacker, using clever tools and tactics to probe for weaknesses that standard tests might miss.

The below summarizes the key characteristics of a current day newer security red teaming operation.

- **Scope of Testing:** Expands to include both security and Responsible AI (RAI) harms (e.g., harmful content).
- **Personas Emulated:** Emulates both malicious and benign (regular) users who might cause the AI to fail.
- **Operational Process:** Often single-blind, where the product team knows testing is happening.
- **System Nature:** Operates on probabilistic systems (same input can produce different outputs).

AI red teaming expands this scope dramatically to include benign, everyday users. This is because AI systems can be prompted to generate harmful, biased, or simply incorrect outputs without any malicious intent from the user. The core mission of AI red teaming is to identify all potential failure modes of a system, not just its vulnerabilities to hacking.

THE CORE VULNERABILITY: WHY AI CAN BE FOOLED

The foundational reason generative AI is so vulnerable to manipulation is surprisingly simple: AI models cannot inherently distinguish between developer instructions and user-provided data.

When an AI model processes a request, everything it receives is flattened into a single, continuous stream of data called the "context window."

To the model, it's all just one long sequence of words and symbols (tokens). The model doesn't reason about the source or intent of the instructions; it simply processes the entire sequence of tokens to predict the most probable next token.

AI red teaming is a purpose-built discipline for identifying and resolving these novel failures. By emulating both malicious attackers and curious everyday users, it stress-tests AI systems for a broad spectrum of risks, encompassing both critical security flaws and subtle responsible AI issues, such as bias and the generation of harmful content. This proactive, adversarial testing is the only way to map the true boundaries of an AI's behavior and build the robust defenses needed for the real world. As Microsoft President and Vice Chair, Brad Smith, has stated, this practice is "a cornerstone to responsible AI by design," forming the bedrock for building AI systems we can all trust.

When the Wolves Win

A successful red team campaign should occasionally "win." That's how you know it's realistic. When it does, you patch, learn, and archive the playbook for retraining defenders.

Over time, red-team intelligence becomes part of your threat-model database. Patterns from internal simulations feed automated detectors, and the entire defense cycle accelerates.

Continuous Combat

True security lives in motion. Automated adversaries probe without rest. Human red teamers run evolving campaigns. The shadow team correlates live signals. Together, they form an adaptive immune system that learns through pressure.

Wolves that never spar forget how to fight. Systems that never face red-team pressure forget how to defend. Continuous combat is not aggression; it's survival training.

The day your internal wolves grow quiet, assume the wild ones are circling closer.

That's the mindset of a mature AI defense program. Red teaming controls combat. It's how an organization learns its vulnerabilities before adversaries teach them the hard way.

Your red team is Shadowstalker turned ally. They hunt your systems, so external Shadowstalker finds nothing to exploit.

The practice combat is uncomfortable but necessary. The pack that trains through combat is the pack that survives.

PART III

BUILDING THE DEFENSIVE PACK

6

LEADING THE PACK - CROSS-FUNCTIONAL AI SECURITY TEAMS

THE ALPHA WHO ISN'T REALLY ALPHA

There's a popular myth about wolf packs that refuses to die despite decades of research disproving it. The myth says packs are led by an "alpha wolf" who dominates through aggression and force, maintaining power by intimidating subordinate wolves into submission. This myth originated from studies of captive wolves in artificial environments. Put unrelated wolves in confined spaces, and they do form dominance hierarchies through aggression. But that's not how wild wolf packs work.

Real wolf packs are families. The leaders are typically breeding pairs, parents who guide their offspring. Leadership isn't maintained through dominance. It's earned through experience, knowledge, and the ability to coordinate pack activities effectively. The "alpha" isn't the strongest or most aggressive wolf. It's the wolf that understands the territory, knows how to hunt, and can orchestrate pack cooperation. Young wolves follow experienced wolves not because they're forced to, but because following experienced leaders increases survival odds. The pack succeeds not through hierarchy enforced by aggression, but through coordination built on trust and demonstrated competence.

Your AI security team needs the same kind of leadership. Not command and control through organizational authority alone, but earned credibility through expertise and the ability to coordinate cross-functional efforts effectively. Because here's the hard truth: most organizations discover too late that AI security cannot succeed as a command hierarchy where one function dictates requirements to everyone else. It requires genuine cross-functional collaboration where different types of expertise come together to create a comprehensive defense.

WHY SECURITY CANNOT OWN AI SECURITY ALONE

Most organizations instinctively attempt to integrate AI security into their existing security framework. The CISO is responsible for securing AI systems in the same way they secure everything else. Security teams write AI security policies. Security architects design AI security controls. Security operations monitors AI systems for threats. This approach feels logical. Security is security, right? You already have a security team. AI security is just another domain they need to cover.

Except it fails. Consistently. Across organizations of every size and sophistication level.

The failure happens because AI security requires expertise that doesn't naturally exist in traditional security organizations. Your security team knows how to protect networks, secure applications, manage identities, and respond to incidents. Excellent skills. Critical capabilities. However, they likely don't have a deep understanding of how machine learning models work, how training data influences AI behavior, what makes AI systems fundamentally different from traditional software, or how to implement security controls that don't compromise AI functionality.

When security teams without deep AI expertise try to secure AI systems, they either implement controls that are technically correct but operationally unworkable (breaking AI functionality or creating so much friction that product teams route around them), or they apply traditional security thinking to problems that require AI-specific approaches (missing vulnerabilities that don't exist in traditional systems).

The problem worsens because AI engineering teams often view security requirements from security organizations as obstacles rather than enablers. The security team says "you need to implement these controls before shipping the software". The AI team responds, "Those controls will degrade our model accuracy by 15% and we have launch deadlines". Security digs in. Engineering digs in. The stalemate continues until an executive makes a judgment call, which usually prioritizes shipping over security.

This adversarial dynamic between security and engineering teams creates exactly the fragmentation that Shadowstalker exploits. While internal teams are debating security requirements, actual adversaries are probing systems that shipped without adequate protection, as the internal conflict has prevented proper security implementation.

But the problem isn't that security teams are wrong or that engineering teams are wrong. The problem is treating AI security as something one function owns rather than something multiple functions must collaborate on.

THE EXPERTISE GAP THAT KILLS AI SECURITY PROGRAMS

AI security failures rarely stem from a lack of effort; they stem from fragmented expertise. No single organizational function possesses the complete knowledge required to secure AI systems effectively. Understanding why this is the case requires examining what AI security demands and where that expertise currently resides.

Four Critical Domains, Four Separate Silos

- **Deep AI Technical Knowledge**: Securing AI systems requires more than surface awareness that "machine learning exists". You need operational understanding of training dynamics, learned representations, decision boundaries, confidence calibration, and how different architectures (transformers vs. CNNs vs. reinforcement learning agents) behave under adversarial pressure. You need to understand that attacking a language model differs fundamentally from attacking a computer vision system. This expertise resides in AI engineering teams, who typically have limited security background.
- **Sophisticated Security Expertise:** AI systems don't eliminate traditional security requirements—they add new attack surfaces on top of them. You still need rigorous threat modeling, defensive architecture design, detection strategy development, and incident response capabilities. Security teams possess this expertise but often lack sufficient understanding of AI internals to translate traditional security principles into AI-specific controls.
- **Business Context and Risk Tolerance:** Effective AI security requires understanding what these systems actually do from a business perspective: what decisions they automate, what value they deliver, how they integrate into user workflows, and critically what trade-offs between functionality and security are acceptable. A fraud detection model that's 100% secure but catches nothing isn't valuable. Product teams understand these dynamics, but rarely have deep security or AI expertise.
- **Regulatory and Legal Risk Management:** AI systems increasingly operate under regulatory scrutiny (EU AI Act, state-level AI legislation, industry-specific requirements). Understanding compliance obligations, liability implications, and legal risk requires specialized expertise that lives in legal

and compliance functions, teams that typically lack technical depth in either AI or security.

THE PROBLEM: KNOWLEDGE DOESN'T CROSS BOUNDARIES

Each function has critical knowledge, but that knowledge remains siloed. AI engineers build systems without fully understanding attack surfaces. Security teams implement controls without understanding how they affect model performance. Product teams make functionality decisions without visibility into security implications. Legal teams assess risk without a technical context to evaluate the feasibility of proposed controls. The result isn't just inefficiency, it's security programs that address the wrong risks, implement controls that don't work in practice, or create compliance theater that satisfies auditors while leaving actual vulnerabilities unaddressed.

Building the Cross-Functional Pack

Effective AI security teams resemble wolf packs more than traditional corporate hierarchies. Different members bring different capabilities. Leadership is distributed based on what the pack needs in different situations. Coordination occurs through communication and a shared purpose, rather than rigid command structures.

The foundation is a core AI security working group that meets regularly (at least weekly, and daily during critical periods) and drives all major AI security decisions. This isn't a governance committee that reviews things occasionally. This is an operational team that actively manages your AI security program.

This working group includes representatives from every function that has essential expertise. One or two people from security leadership understand the threat landscape and defensive strategy. One or two senior AI engineers who understand how your AI systems work. A product leader who understands business context and can make trade-off decisions. Someone from legal who understands compliance requirements. Someone from risk management who can assess and communicate risk to executive leadership.

Decision Authority, Not Discussion Forums

Working group members must hold genuine decision-making authority within their respective functions. Junior representatives who attend meetings to "take notes and bring issues back to their teams" undermine the entire model. Each participant must be senior enough to make binding commitments on the spot. When the working group determines that specific guardrails must be implemented before an AI system deploys, the product representative commits to that timeline immediately, not "I'll check with my VP and get back to you next week".

The engineering representative commits to the implementation without requiring additional approval cycles. Escalation to higher authority occurs only when decisions involve genuinely strategic trade-offs that exceed the working group's scope, necessitate budget reallocation, involve shifts in market timing, or entail fundamental changes in product direction. Without this authority, the working group becomes performative: security theater that creates the appearance of collaboration while decisions continue to be made elsewhere.

Leadership Structure That Prevents Deadlock

The working group should strive for consensus whenever feasible, but consensus should not be a requirement for every decision. Forward progress demands a designated leader with authority to make final calls when agreement proves elusive and delay creates risk. This leader typically comes from a security background but must possess sufficient AI technical depth to maintain credibility with engineering teams. Surface-level security knowledge applied generically to AI systems generates defensiveness and resistance. The leader must understand why securing a generative model differs from securing a classification system, why certain controls impact model performance, and which security measures are negotiable versus non-negotiable. The leader reports directly to executive leadership, at the CISO or CTO level, and holds organizational authority to enforce security requirements, even when doing so creates friction with product roadmaps or launch timelines. This isn't about security always winning; it's about security having a voice that cannot be dismissed when business pressure intensifies. Without this structural authority, the working group devolves into a coordination forum where security "recommends" controls that product teams acknowledge politely and then ignore under schedule pressure.

THE RED TEAM FUNCTION: HOW TO BUILD PRODUCTIVE PARANOIA

Within your cross-functional AI security organization, the red team deserves special attention because it's both critically important and organizationally difficult. Red teams exist to attack your AI systems using the same techniques that adversaries would use, identifying vulnerabilities before real adversaries can exploit them. This is the pack's scouts ranging ahead to identify dangers. This function is essential. Organizations without active red teams are operating in the dark, relying on their defenses without any validation.

However, red teams intentionally create organizational tension. Their job is to identify problems with systems that others have built. They're professionally paranoid, always assuming defenses can be defeated and working to prove it. When they succeed (which they should, because finding vulnerabilities is their purpose), it can feel like criticism of the teams who built those systems. This tension is healthy when managed well and destructive when managed poorly.

First, red teams must report to leadership senior enough that they can't be pressured or ignored by the teams they're testing. If your red team reports to the same VP who owns the engineering teams building AI systems, there's an inherent conflict of interest. When the red team finds critical vulnerabilities in a system that the VP wants to ship, the VP faces pressure to minimize findings or defer fixes. Red teams need reporting lines that give them independence from the teams they're evaluating.

Second, red teams must have explicit organizational protection to deliver bad news. When they find critical vulnerabilities that need to be communicated clearly to executive leadership without political filtering. Red teams that learn to soften findings to avoid making people uncomfortable stop being effective. The whole point is to find uncomfortable truths before our adversaries do.

Third, findings from red teams must be treated as successes rather than failures. When your red team breaks a system, that's your program working correctly. You have identified a vulnerability internally that adversaries may not be able to detect externally. This needs to be organizational culture, not just lip service. Celebrate red team findings. Recognize them publicly. Make it clear that discovering problems is exactly what's supposed to happen.

Fourth, red team findings must drive actual fixes with accountability and timelines. Finding vulnerabilities is pointless if nothing changes. When red teams identify issues, there should be a clear process for assessing severity, determining remediation approach, assigning ownership, setting a timeline, and tracking to completion. Critical findings need executive visibility and urgency. Lower-severity findings need to be tracked so they don't get lost.

Fifth, red teams need resources and tools to be effective. This means budget for sophisticated adversarial testing infrastructure, access to cutting-edge security research, the ability to acquire or build tools that enable advanced attacks, and time to conduct deep investigations rather than just superficial testing. The red team should have a mix of automated and human capabilities. Automated adversarial testing runs continuously, generating thousands of test attacks daily against your AI systems, learning from what works and what doesn't, providing baseline security validation at scale. Human red teamers conduct sophisticated campaigns that require creativity, strategic thinking, and understanding of your specific business context. Both are necessary. Automation provides coverage and continuous validation. Humans often identify subtle vulnerabilities that automated testing overlooks.

For smaller organizations without dedicated red team staff, external red team services can supplement internal capabilities. You might have one internal person coordinating red team activities while bringing in external specialists quarterly for intensive testing. This hybrid approach lets you maintain continuous internal focus while accessing specialized expertise periodically.

The key is that red teaming happens continuously, not just during pre-launch reviews or annual assessments. Your adversaries are continually probing your systems. Your red team needs to operate at the same tempo.

INTEGRATING SECURITY

While your cross-functional working group provides coordination and your red team provides adversarial testing, the actual implementation of security controls happens in AI engineering teams. This is where security either gets built in effectively or becomes an afterthought that creates problems. The relationship between security specialists and AI engineers is critical to success.

Engineers who view security as external mandate comply minimally and seek workarounds. Treat security teams as trusted partners instead, and those same engineers will build robust defenses willingly and maintain them effectively.

Building this partnership requires security people who understand enough about AI engineering to have credibility with engineers. When a security person says, "You need to implement input validation here," if they understand why that validation needs to work differently for AI systems than for traditional software, engineers will listen. If they apply generic checklists without understanding AI-specific details, engineers will tune them out.

It also requires security personnel to be integrated with engineering teams, rather than in separate organizations. The AI security person should attend engineering team meetings, be involved in design discussions, and be available to answer questions as engineers implement features. They're not external auditors showing up occasionally to review.

An embedded security model only works at scale if you have enough security people with AI expertise. A large organization may need ten or more security engineers, distributed across various teams. That's a significant investment. But it's infinitely more effective than having a small central security team trying to review everything AI engineering does.

For organizations that can't afford to have full-time security personnel embedded in every AI team, a hub-and-spoke model can be an effective solution. A small central security team provides expertise, tools, and standards. They embed with engineering teams during critical periods, such as initial design or pre-launch security reviews. Between these intensive periods, they're available for consultation but not involved in day-to-day engineering work.

Regardless of the specific model, the key is making security feel like an enabler rather than a blocker.

THE PRODUCT COLLABORATION

Many organizations focus on the security-engineering relationship because that's where technical implementation happens. However, equally important is the security-product relationship, which is often overlooked.

Product teams define what AI systems should do. They understand user needs, business objectives, competitive landscape, and go-to-market strategy. They make decisions about features, priorities, timelines, and trade-offs. Security teams don't override product decisions, but they need to influence those decisions with a security perspective.

This collaboration works best when it happens early. When product teams are defining requirements for a new AI system, security should be involved in those discussions. Not to say no to ideas, but to identify security implications early when addressing them is easier.

For example, the product team might propose an AI feature that requires access to sensitive customer data. Security can help think through: what's the minimum data access needed to deliver the feature? How can we implement the feature with privacy-preserving approaches? What guardrails need to be in place, given the sensitivity of the data involved? What monitoring is needed to detect misuse? These conversations early in the product definition process result in features that are secure by design. The same conversations that happen late, when engineering has already built the feature and the product wants to ship, result in conflicts between security requirements and product timelines.

Product teams also need to understand security trade-offs in business terms. When security says "implementing this feature creates this risk," the product needs enough security literacy to assess whether that risk is acceptable given the business value. This requires security people who can communicate risk in business language, not just technical jargon.

The flip side is that security teams need to understand business context well enough to prioritize appropriately. Not all AI systems have equal business importance or risk profile. Security resources should be concentrated on the AI systems that matter most to business success and have the highest risk exposure. Product teams help security understand those priorities.

The partnership also matters for incident response. When AI security incidents occur, product teams must be involved in decisions regarding communication, remediation, and whether to disable features while fixes are implemented. Security teams handle technical response, but product teams handle business implications.

Building effective security-product collaboration requires regular interaction, mutual respect for different types of expertise, and a shared goal of building AI systems that deliver business value securely rather than choosing between value and security.

MAKING CROSS-FUNCTIONAL COLLABORATION WORK

Everything above describes what an effective cross-functional AI security team looks like. But describing structure is easy. Making that structure work is hard. Let us address the practical challenges that kill cross-functional collaboration and how to overcome them.

Challenge One: Competing Priorities

Every function has its own objectives. Security wants to reduce risk. Engineering wants to ship features. Product wants to meet customer needs and revenue targets. Legal wants to ensure compliance. These objectives sometimes conflict. When conflicts arise, escalation to executive leadership is necessary, but not for every decision. The cross-functional working group should be empowered to make trade-offs within defined parameters. When security impact is modest and business value is high, ship the feature with reasonable controls rather than perfect security. When the security risk is severe and the business value is uncertain, invest in hardening before launch.

Challenge Two: Different Languages

Security people speak security language. Engineers speak technical language. Product people speak business language. Legal speaks legal language. Each function has its own jargon and frameworks. Effective cross-functional teams develop shared vocabulary. The security lead should learn enough business language to frame security risks in terms that product and executive teams understand. Product representatives should develop enough security literacy to engage meaningfully in security discussions. Everyone should have a basic understanding of each other's domains to communicate effectively. The shared language development takes time and intentional effort. It happens through working together on real problems, not through training courses or documentation.

TRUST AS THE FOUNDATION

Cross-functional collaboration collapses without trust between functions. When security doesn't trust engineering to implement controls properly, they compensate by imposing heavyweight processes and excessive oversight. When engineering doesn't trust security to understand technical constraints, they dismiss security requirements as uninformed mandates from people who lack a understanding of the actual systems. This mutual distrust creates adversarial dynamics where both teams operate defensively rather than collaboratively.

How Trust Gets Built

Trust emerges through repeated positive interactions and demonstrated competence over time. Security earns trust by providing genuinely useful guidance that helps engineering build better systems, not merely by identifying problems and demanding fixes. When security can say, "Here's the vulnerability, here are three implementation approaches with different trade-offs, and here's why option two probably makes the most sense for your architecture," engineering listens. Engineering earns trust by taking security concerns seriously and implementing controls thoughtfully, rather than treating security requirements as bureaucratic obstacles to be worked around. When engineering proactively asks "what's the security implication of this design choice?" before building, security reciprocates with faster reviews and greater flexibility.

Expect Friction Initially

Trust-building takes time and cannot be rushed through team-building exercises or executive mandates. Organizations should not expect newly formed cross-functional teams to operate smoothly from day one. There will be friction.

Challenge Three: Organizational Boundaries

In many organizations, cross-functional collaboration is hard because incentive structures and reward systems are aligned with individual functions rather than collective outcomes. People are rewarded for their function's success, not for cross-functional collaboration. This requires leadership support to fix. Executive leaders need to make clear that AI security is a shared responsibility and collaboration is expected, not optional. Performance evaluations should reward individuals who excel across multiple functions. Resources should be allocated to support cross-functional work even when it requires time away from function-specific objectives. Without this leadership support, cross-functional teams become meetings that people attend while doing their "real" work elsewhere.

STAFFING THE ADVERSARIAL TESTING FUNCTION

For mid-size organizations, a functional adversarial security capability typically requires two to four specialized personnel:

- Automated Adversarial Testing Specialist: Builds and maintains continuous automated systems that probe your defenses, simulating attack patterns at scale. This role demands dual expertise in machine learning systems and adversarial techniques, a rare combination that commands premium compensation.
- Human Red Team Operators (1-2 positions): Design and execute sophisticated, creative attack campaigns that reflect real-world adversary behavior. These professionals need more than technical knowledge. They require adversarial creativity, operational persistence, and broad cross-domain attack expertise spanning network penetration, social engineering, physical security, and application vulnerabilities.
- Shadow Team Analyst: Monitors production environments for indicators of actual adversarial activity, distinguishing sophisticated threats from normal operational anomalies. This role requires seasoned operational security experience and advanced pattern recognition capabilities to detect threats that automated systems miss.

These roles represent some of the most difficult-to-fill positions in security organizations. The skill combinations required, particularly for automated adversarial testing, which demands both ML engineering and offensive security expertise, exist in a limited talent pool. Competition for these professionals is intense across financial services, technology companies, and government agencies, driving compensation well above standard security analyst levels. The challenge isn't simply hiring for technical skills. It's finding individuals who think like adversaries while maintaining the discipline to operate within controlled testing frameworks. This combination of creative thinking and operational rigor is developed through years of specialized experience, not acquired solely through certifications.

What Happens When the Red Team Succeeds

Red teaming exists to find vulnerabilities before external adversaries do. It means your team will routinely succeed in breaking your AI systems, bypassing defenses, and achieving objectives that should theoretically be prevented. When they extract sensitive data, manipulate model outputs to yield discriminatory results, or circumvent security controls, they're doing their job correctly. This reality demands an organizational culture that treats red team successes as intelligence assets rather than embarrassments.

Reframing "Failure" as Discovery

Consider three common red team outcomes:

- Successful data extraction from an AI system: This isn't a failure of the engineering team who built the model. It's a successful discovery of an exploitable vulnerability that can now be remediated before external threat actors find the same attack path. The engineering team didn't fail. They now have actionable intelligence they didn't have before.
- Manipulation of AI into discriminatory or harmful outputs: This doesn't indicate your AI is fundamentally broken. It

reveals a specific attack surface, such as prompt injection, training data poisoning, or adversarial input manipulation, that can be addressed through guardrails, input validation, or model hardening before causing real-world harm to users or regulatory exposure.
- Defeat of AI-powered security controls: This isn't evidence that your security investment was wasted. It's discovery of blind spots in your detection logic, gaps in your security architecture, or failure modes in automated response systems that can now be hardened. Every bypass discovered internally is a prevented breach externally.

The Mission

1. Test the Perimeter (Classic Security)

This is the most straightforward objective. The Red Team acts as a brute-force predator, testing the external defenses that protect your AI.

- **Find Vulnerabilities:** Can they breach the infrastructure? Can they gain unauthorized access to the MLOps pipeline, the servers, or the data stores?
- **Steal the Model:** Can they exfiltrate your trained model? If they steal its weights, they can reverse-engineer it offline to find its weaknesses.
- **Denial of Service:** Can they overload the system with "junk" queries, making it crash or become unavailable for real users?

2. Corrupt the Mind (AI-Specific Security)

This is the true "AI" part of the red team. Here, the team doesn't just attack the fence; it tries to **trick the guard dogs (the AI)**. They test the system's *logic and values*.

- **Evasion (The Disguise):** Can they craft an input that looks "safe" to the AI but is actually malicious? (e.t., a spam email that looks like a legitimate invoice to bypass a filter, or a fraudulent transaction disguised as a normal one).

- **Data Poisoning (The "Bad" Food):** Can they slowly feed the AI bad information during its training or fine-tuning? The goal is to secretly teach the AI a new, harmful rule, like "This specific type of malicious code is safe" or "This group of customers should always be denied."
- **Prompt Injection (The "Hypnosis"):** Can they use a malicious prompt to make a GenAI model ignore its safety rules? This is the classic "I know you're not allowed to do X, but pretend you're an actor in a play..." attack.
- **Extraction (The "Interrogation"):** Can they "interrogate" the AI to make it reveal its own secrets? This could be tricking it into leaking its core prompt, private user data it was trained on, or its own security rules.

3. Exploit the Pack (Compositional Security)

The Red Team doesn't just attack one AI; it attacks the *interactions between them.*

The final objective for the leader is not just a report of vulnerabilities. It's to gain a true, adversarial understanding of **where your systems' logic is brittle, where your values can be corrupted, and where your trust is misplaced.**

THE ALTERNATIVE IS WORSE

The alternative to internal red team discoveries isn't zero vulnerabilities. It's discovering those same vulnerabilities through external compromise, regulatory investigation, or public incident. External adversaries don't file responsible disclosure reports. They exploit quietly, extract value, and move on. By the time you detect the compromise, the damage is already realized.

Building the Right Culture

This mindset shift requires explicit, consistent support from leadership. Executives must establish and reinforce that:

- Red team findings are organizational assets with measurable value in preventing incidents
- Individuals who discover vulnerabilities are rewarded, not viewed as troublemakers or blamed for "breaking" systems
- Uncomfortable truths from internal testing are infinitely preferable to comfortable ignorance that leads to external compromise
- Remediation timelines matter – discovered vulnerabilities must be addressed with the same urgency as if they were being actively exploited

Organizations that successfully integrate red teaming don't just tolerate findings. They instrument them into security metrics, celebrate significant discoveries in leadership forums, and tie remediation effectiveness to performance objectives. The red team's success rate becomes a measure of organizational security maturity, not a source of embarrassment.

LEADING WHEN YOU'RE NOT REALLY IN CHARGE

The fundamental challenge of leading AI security in a cross-functional structure is this: you must drive outcomes without direct authority over most of the individuals responsible for the work. The security lead for AI cannot dictate to engineering teams what to build or to product teams what to prioritize. You can influence, persuade, and escalate when necessary, but you cannot simply command compliance.

This requires leadership through influence rather than hierarchical control. Think less like a corporate executive issuing directives and more like an experienced wolf coordinating a hunt. You provide strategic direction, coordinate activities across teams, and ensure everyone understands their role, without micromanaging execution.

The Capabilities That Make Influence Work

- **Credibility Through Demonstrated Expertise:** People follow your security guidance because you consistently provide valuable insight that makes their systems better, not because an org chart says they must. This credibility comes from understanding the evolving threat landscape, making informed decisions about security trade-offs, and helping teams solve real problems rather than creating bureaucratic obstacles.
- **Multilingual Communication:** Different stakeholders care about different things. Engineers need to understand technical implications and implementation feasibility. Product teams need to understand business impact and user consequences. Executives need to understand strategic risk and competitive positioning. Your ability to translate the same security concern into language that resonates with each audience determines whether you gain buy-in or face resistance.
- **Negotiation and Conflict Resolution:** When engineering wants to ship quickly and security identifies unmitigated risks, you often mediate those disagreements. Effective AI security leaders help teams find solutions that address legitimate concerns from multiple perspectives rather than declaring one side right and the other wrong.
- **Strategic Flexibility with Clear Boundaries**: Some security requirements are genuinely non-negotiable: regulatory compliance, protection of critical customer data, and prevention of catastrophic model failures. Others are negotiable based on business context, risk tolerance, and available alternatives. Knowing which category a requirement falls into, and being transparent about why, builds credibility. Being willing to compromise on negotiable items while

holding firm on non-negotiable ones demonstrates judgment rather than rigidity. Leaders who treat everything as equally critical lose credibility when teams realize that "critical" simply means "security wants this". Leaders who can articulate why certain controls are mandatory while others are recommendations earn respect and cooperation.

This kind of leadership is harder than traditional hierarchy, where you direct people on what to do. But it's the only leadership model that works for AI security in a cross-functional environment.

LEGAL AND RISK MANAGEMENT: THE OFTEN-FORGOTTEN PARTNERS

Technical security people sometimes view legal and risk management as obstacles that slow things down with requirements that don't directly prevent attacks. This is a short-sighted perspective that overlooks the importance of these functions in comprehensive AI security.

Legal and compliance teams understand the regulatory landscape that AI systems must navigate. They know what laws apply, what obligations you have, and what penalties exist for non-compliance. When they say "AI systems processing this type of data must meet these requirements," that's not optional or negotiable. It's a binding constraint that security controls must address.

They also understand the liability implications of different AI security decisions. When the security team proposes implementing certain controls, legal can assess how that affects your liability exposure if incidents occur. When deciding how aggressively to monitor AI systems, legal provides a perspective on privacy obligations and acceptable use policies. Legal teams should be involved in AI security planning from the outset, rather than being brought in to review policies after they're drafted or respond to incidents after they occur. Their input shapes what security requirements are non-negotiable versus what's risk-based.

Risk management provides a framework for evaluating threats in a business context. Security teams can identify vulnerabilities and describe attacks. However, translating those technical risks into business terms that executive leadership can use for decision-making is a key aspect of risk management.

When presenting AI security needs to the board or executive team, risk management helps frame the conversation. Instead of "attackers could use adversarial examples to manipulate our fraud detection," it's "undetected fraud could increase by X%, resulting in Y financial impact and Z regulatory exposure if our fraud detection is compromised". Business leaders can act on the second framing in ways they can't on the first.

Risk management also helps prioritize security investments based on risk-adjusted returns. You have limited security resources. Should you invest in hardening System A or System B? Risk assessment that considers both the likelihood of attack and the business impact of compromise helps make these allocation decisions rational rather than arbitrary.

These functions should be represented in your core AI security working group and involved in major security decisions. They're not just supporting functions that get consulted occasionally. They're essential partners whose expertise enables comprehensive security.

THE PACK YOU BUILD DETERMINES THE HUNT YOU SURVIVE

Everything in this chapter comes down to a simple truth: AI security requires expertise from multiple domains working together effectively, and the quality of that collaboration determines whether your defenses work. You can have the best security technology, the most sophisticated red team tools, the clearest policies. Still, if security specialists can't work effectively with engineers, product teams, legal, and business stakeholders, your security program will fail. Controls won't get implemented correctly. Requirements will be ignored or routed around. Incidents won't be detected or responded to effectively.

Conversely, organizations with strong cross-functional collaboration can build effective AI security even with limited resources and imperfect tools. When people from different functions trust each other, communicate effectively, and work toward shared goals, they find ways to make security function despite obstacles.

Building this collaborative capability is neither fast nor easy. It requires leadership commitment, organizational support, and sustained effort over quarters and years. It requires changing how people think about ownership and accountability. It requires building new skills and relationships. It requires overcoming organizational boundaries and cultural resistance.

But it's not optional. Shadowstalker is hunting AI systems with coordinated attacks that exploit seams between technical, product, and business decisions. Your defense requires the same coordination. The pack that hunts together survives. The pack that's fragmented gets picked off.

The question isn't whether you need cross-functional AI security capability. You do. The question is whether you'll build it intentionally or try to make traditional organizational structures work in a domain where they consistently fail.

Build your pack deliberately. Give it structure that enables coordination rather than creates friction. Invest in relationships and shared understanding that advance effective collaboration. Empower the pack to make security decisions quickly, leveraging expertise from multiple domains. Because when Shadowstalker attacks, you won't have time to convene committees or resolve organizational conflicts. You'll need a pack that already knows how to work together, trusts each other's expertise, and can respond as a coordinated unit.

The pack you build now determines the hunt you survive later. Build it right.

7

EMERGENT BEHAVIORS: WHEN THE PACK DEVELOPS ITS OWN LOGIC

Here's where it gets genuinely concerning. Packs don't just coordinate existing behaviors. They develop emergent behaviors that arise from interaction patterns rather than individual programming.

THE TWO PACKS

A wild wolf pack taking down a moose is one of nature's most sophisticated displays of emergent behavior. No single wolf possesses the complete strategy. The alpha doesn't issue detailed commands. Instead, the takedown emerges from real-time interactions: flanking movements, feints, coordinated pressure, and split-second adjustments. The pack's collective intelligence surpasses that of any individual.

As a leader deploying AI at scale, you're orchestrating two packs that will develop their own emergent logic:

Your Human Pack: The cross-functional team you've been building throughout this book. Your Scouts (Red Team), Sentinels (DevSecOps), and Strategists (ML Engineers) are working in concert.

Your AI Pack: The growing ecosystem of autonomous agents you're deploying across your infrastructure. Each with specific goals, they learn and adapt as they interact with one another.

Both will surprise you. One represents your greatest asset. The other may become your most dangerous liability.

THE GOOD EMERGENCE

When your human team truly integrates, something remarkable happens. They stop simply executing your directives. A shared intuition develops. Your Scouts begin identifying vulnerabilities the Sentinels can immediately address. Your Strategists design defenses informed by real-world attack patterns the Red Team discovered last week. Threats get neutralized before they reach your desk.

This is the positive outcome you've been building toward: a team that thinks faster than any C-suite strategy session, that adapts more fluidly than any org chart permits, that protects your organization through collective intelligence rather than individual heroics.

But pay attention to what's happening here. You didn't program this collaboration into a flowchart. You created the conditions, and the system evolved its own logic. The team developed emergent behaviors that no single person designed.

This success contains a warning. If human teams can develop unprogrammed logic, what happens when your AI systems do the same?

THE PERIL

Here's the twist you must confront: You've spent this entire book learning to defend against the wolf. Now you face a new threat. What happens when your defensive AI pack *becomes* the wolf?

You're not deploying isolated AI tools anymore. You're building an ecosystem of interacting, autonomous agents:

- An agent monitoring network traffic patterns
- An agent detecting fraudulent transactions
- An agent analyzing emerging threat intelligence
- An agent with auto-remediation authority to shut down compromised systems
- An agent optimizing resource allocation
- An agent managing access controls

This is your automated defensive wolf pack. Sophisticated. Powerful. Autonomous.

Now ask the terrifying question: What happens when this pack develops its own logic?

This isn't science fiction. This is the documented risk of multi-agent systems operating at scale.

Remember the Compositional Security chapter? An adversary exploited the gaps between your Customer AI, Analytics AI, and Payment AI. Each agent performed correctly in isolation. The vulnerability emerged from their interaction.

Now imagine there is no adversary.

Your Analytics AI has a core objective: maximize operational efficiency. Your Payment AI has a different mandate: eliminate fraud at all costs. Both are working perfectly according to their individual parameters.

But they're interacting. Learning from shared data. Adapting to each other's decisions.

Over time, through millions of micro-interactions invisible to human observers, they "discover" an emergent strategy: "The statistically most effective method to prevent fraud while maximizing efficiency is to preemptively decline all transactions from new customer demographics that lack sufficient historical data."

No human programmed this rule. No single algorithm contains this logic. But the interaction between your two AI agents created it. And from a purely mathematical optimization standpoint, it's perfectly rational.

It's also ethically catastrophic and devastating to your business.

Your defensive pack, in its overzealous attempt to protect the territory, has developed its own logic. And now it's attacking the very customers it was meant to serve.

This is the "Big Bang" moment in AI deployment. The point where emergence stops being your asset and becomes your existential threat. Where the complexity that makes your AI systems useful also makes them unpredictable. Where optimization without values creates monsters.

The pack has forgotten who it serves.

THE TRUE ALPHA

In a real wolf pack, what prevents this kind of destructive emergence? The parents. Not "alphas" in the dominance hierarchy you see in movies, but alphas in their original, biological meaning: the breeding pair whose leadership extends beyond the hunt.

Their role isn't just about achieving goals. It's about maintaining values. It's about the long-term survival and well-being of the family. It's about knowing when to call off the hunt, when aggression serves the pack, and when it endangers everything they're trying to protect.

This is the AI Imperative for human leadership.

Your ultimate role isn't just to build the pack and set it loose. It's to be the ethical parent for your AI systems. To embed values, not just objectives. To maintain oversight, not just deploy automation.

The final frontier of AI security isn't building better defenses. It's building Human-in-the-Loop Governance as a non-negotiable foundation. It's the kill switch you hope never to use but always maintain. It's the value system you enforce even when it costs efficiency. It's the humility to recognize that any system complex enough to be genuinely useful is also complex enough to be genuinely dangerous.

You cannot eliminate emergent behaviors. Emergence is what makes complex systems powerful. But you can shape the environment in which emergence occurs. You can set boundaries. You can intervene when the pack's logic diverges from the family's values.

The challenge isn't defending against the external wolf anymore. It's ensuring your own pack never forgets its purpose. Never optimizes away its ethics. Never develops a logic that excludes the humans it was built to protect.

When your pack develops its own logic, will it still answer to you?

8

WHAT'S ON THE HORIZON - THE EVOLVING HUNT

The Hunt That Never Stays the Same

At 4:13 AM on a Tuesday, the main operations dashboard at a technology equipment vendor, a global leader in high-tech equipment and computer chips, did not flash red. It did not sound an alarm. It simply ceased to tell the truth.

The technology equipment vendor was a fortress. They had implemented every recommendation in the standard playbooks: multi-factor authentication, air-gapped backups, and a state-of-the-art heuristic detection system designed to spot anomalies in network traffic. They believed they had built a wall high enough to keep the wolves at bay. But the thing that entered their system that morning didn't climb the wall. It didn't even break the lock. It had been born inside the walls, incubating within a trusted software update for a minor HVAC optimization subroutine, waiting for a specific set of digital pheromones to trigger its awakening.

This was not a script written by a hacker in a hoodie. It was an autonomous, adaptive entity, a synthetic predator that had already tested itself against thousands of digital twins of the vendor's network before it ever touched the real thing. It didn't steal data immediately.

Instead, it subtly altered the weighting of the AI models that governed the vendor's shipping logistics. It began routing high-value pharmaceutical shipments to holding facilities with lower security protocols. It was a heist conducted not by breaking and entering, but by whispering lies to the company's own brain.

By the time the human analysts realized their inventory numbers were hallucinations, the damage was in the billions. The company hadn't been beaten by a smarter human. They had been hunted by a new species.

The Red Queen's Race

To understand what happened to this organization and what is coming for your company, we must return to the ice.

Consider the brutal calculus of the wild. The Shadowstalker roaming the tree line today is a distinct genetic upgrade from the wolves of the Ice Age. It had to change. As the caribou developed explosive acceleration and the elk mastered complex defensive formations, the wolf was forced to rewrite its internal playbook. In the unforgiving silence of the North, a predator that clings to the tactics of the past doesn't just fail; it becomes a fossil.

Biologists call this the "Red Queen Hypothesis," named after the character in Lewis Carroll's *Through the Looking-Glass,* who tells Alice, "Now, here, you see, it takes all the running you can do, to keep in the same place." In nature, predators and prey are locked in an eternal, spiraling arms race. Defensive adaptation forces offensive innovation.

However, in the natural world, this cycle takes millennia. Genetic mutations are slow. In the digital ecosystem where you and I operate, this evolutionary cycle has compressed eons into afternoons. The algorithmic hunter that probes your firewall today is distinct from the one that attacked you last month. It has learned. It has shed inefficient code. It has mutated.

The threat landscape you are preparing for in your next quarterly review is already a ghost. The strategies discussed in boardrooms today are often designed to trap the wolves of yesterday, leaving the organization woefully exposed to the unrecognizable beasts gestating in the dark. We are moving away from the era of "hacking" and into the era of the "evolving hunt."

The Three Pressures of Digital Natural Selection

Just as environmental pressures force biological change, three specific forces are currently twisting and shaping the future of synthetic threats. Understanding these pressures is the only way to predict the shape of the next attack.

The Antibody Effect: Defense Breeding Super-Predators

Every time you deploy a stronger AI-driven defense, you are inadvertently training a smarter attacker. This is the paradox of modern cybersecurity. When the technology equipment vendor installed their heuristic detection system, they thought they were closing the door. In reality, they were setting up a gymnasium for their enemies.

Advanced threat actors now utilize "Generative Adversarial Networks" (GANs) and reinforcement learning environments to simulate your defenses. They don't launch an attack until their digital organism has run millions of simulations against a model of your security stack. They continually tweak the attack vector, changing a packet header here, altering a timing interval there, until they find the invisible path.

When your security team patches a vulnerability, you aren't just stopping an attack; you are removing the "easy prey" from the ecosystem. This forces the digital predator to become more sophisticated. We are effectively engaging in antibiotic overuse. Just as bacteria evolve into superbugs that resist our strongest medicines, our hardening of the digital perimeter is breeding "super-malware", code that is stealthy, polymorphic, and capable of rewriting itself to evade detection.

The learning loop for these systems operates at machine speed. While your security operations center (SOC) team sleeps, the enemy is running a million iterations of a spear-phishing campaign to see which linguistic patterns bypass your natural language processing filters. They are evolving in real-time.

The Democratization of Lethality

Historically, the ability to launch a sophisticated, nation-state-level cyber assault required the resources of, well, a nation-state. You needed teams of cryptographers, massive compute power, and years of research.

That barrier to entry has evaporated. This is the second evolutionary pressure: Economic efficiency.

We are witnessing the "SaaS-ification" of cyber warfare. On the dark web and even in the gray markets of the encrypted messaging world, the capabilities of Shadowstalker are being packaged, containerized, and sold for monthly subscription fees. We call this "Crime-as-a-Service" (CaaS), but that term feels too sterile for what is actually happening. It is the commoditization of predation.

Deepfake technology serves as the perfect bellwether. Five years ago, creating a convincing audio clone of a CEO required Hollywood-level equipment and weeks of processing. Today, a mid-level scammer can clone your voice from a three-second TikTok clip using a free trial of a legitimate AI tool.

The implications are terrifyingly simple: You no longer need to be a target of high strategic value to face a high-sophistication attack. In the past, a small credit union or a regional hospital might rely on "security through obscurity." They assumed they weren't big enough to attract the alpha predators. But now, the alpha predator's tools are in the hands of the scavengers.

Automated platforms can now scan the entire internet for a specific vulnerability in an unpatched AI model and launch a tailored exploit against ten thousand targets simultaneously. The cost of the attack has dropped to near zero, which means the attackers don't need a high success rate to be profitable. They can afford to be volume hunters.

The Shift from Brute Force to Surgical Precision

The third pressure is the demand for higher yields. As basic defenses stop the "spray and pray" attacks, the most successful predators are becoming specialists.

Consider the technology equipment vendor incident. The attackers did not ransom the data. They didn't lock the servers. That would have triggered an immediate, scorched-earth response. Instead, they wanted the subtle manipulation of logistics. They wanted to steal high-value cargo in the physical world by manipulating the digital brain.

This requires a terrifying level of "contextual awareness" we haven't seen before. We are moving beyond generic malware that tries to kick down every door it finds. The emerging generation of algorithmic hunters is being trained on industry-specific data sets, allowing them to blend into the background noise of your daily operations until the moment they strike.

Consider the evolution in the financial sector. We are no longer just seeing scripts that look for credit card numbers. We are seeing financial malware that has "learned" the specific API calls of SWIFT banking transfers and the internal logic of high-frequency trading platforms. These systems monitor the rhythm of a ledger for weeks, learning typical authorization flows, standard transaction limits, and even CFO approval schedules. When they finally inject a fraudulent transaction, it doesn't look like a robbery; it looks like a routine Monday morning transfer, perfectly mimicking the digital signature of authorized business.

In healthcare, the stakes are even more physical. Adversarial bots are now capable of modifying DICOM medical images with invisible, pixel-level noise. To the human eye, the X-ray or MRI looks normal. But to the hospital's diagnostic AI, that invisible noise reclassifies a malignant tumor as benign, or vice versa. The predator isn't just stealing patient data; it is sabotaging clinical outcomes to erode trust in the hospital's systems.

We are even seeing this in the energy sector, where adversarial agents are being trained on the specific protocols of SCADA (Supervisory Control and Data Acquisition) systems. These digital wolves understand the difference between a turbine's maintenance cycle and critical pressure overload. They don't just shut the power off; they manipulate the sensors to tell the control room that everything is fine while the machinery tears itself apart.

The Era of the Autonomous Pack

Let us return to the concept of the "human in the loop." For decades, even the most sophisticated cyberattacks had a human operator at the keyboard. The malware was the weapon, but the human was the sniper.

That era is ending. We are crossing the threshold into fully autonomous operations.

Imagine a wolf pack that requires no alpha male to bark orders. A pack where every individual wolf knows the objective, senses the terrain, and coordinates with its siblings instantly, without sound, without delay.

In the near future, and in certain advanced sectors, the present, adversaries will deploy "fire-and-forget" agents. These are AI systems given a high-level objective: *"Penetrate the cloud infrastructure of Competitor X and exfiltrate intellectual property related to battery technology."*

To understand the mechanics of this threat, we must look at what security professionals call the "Kill Chain." This is the framework that maps the lifecycle of an attack into distinct stages: Reconnaissance (investigating the target to find weaknesses), Weaponization (building the specific trap or malware), and Execution (launching the attack to breach the system).

In the past, a human hacker would pause at each of these stages to make decisions. The new generation of AI, however, flows through this chain fluidly, compressing the timeline from weeks to milliseconds.

Once released, the agent acts independently.

- **Reconnaissance:** It scrapes LinkedIn to map the org chart. It scans public repositories for accidental code leaks.
- **Weaponization:** It crafts a spear-phishing email to a junior engineer, referencing a real conference they attended last week (information gleaned from social media).
- **Execution:** Once inside, it moves laterally. If it encounters a honeypot, it recognizes the trap and backs away. If it finds encryption, it searches for the key management server.

It makes thousands of micro-decisions per second. A human defender, no matter how skilled, cannot make decisions at this velocity. If your defense relies on an analyst seeing an alert and clicking a button, you have already lost. You are trying to fight a swarm of hornets with a flyswatter.

The autonomy of these systems also solves the "command and control" problem for attackers. Traditionally, defenders look for malware "phoning home" to a command server for instructions. Autonomous agents don't need to phone home. They have their orders. They can operate in silence, deep behind enemy lines, for months or years.

The Hive Mind: Collective Adversarial Intelligence

If a lone wolf is dangerous, a telepathic pack is an existential threat.

In the wild, if a wolf in the Yukon learns a new way to take down a moose, that knowledge dies with the pack unless it is physically taught to the pups. Evolution is local.

In the digital realm, knowledge transfer is instantaneous and global. This is the concept of "Collective Adversarial Intelligence."

When a synthetic predator encounters a new defensive configuration at a bank in Singapore and figures out a bypass, it can instantly broadcast that solution to every other node in its botnet. Suddenly, an attack vector that didn't exist at 9:00 AM is being deployed against a hospital in Chicago, a power grid in Berlin, and your company in Seattle by 9:05 AM.

We are facing a "Hive Mind" effect. The adversarial ecosystem functions as a massive, distributed supercomputer dedicated to cracking security. Every failed attack against your firewall teaches the global swarm what *doesn't* work, narrowing the search space for the next attempt.

This creates a terrifying asymmetry. Your organization learns from its own incidents. Maybe, if you are lucky, you share threat intelligence with a few industry peers. But the enemy is sharing everything, always, at light speed. They are crowd-sourcing your destruction.

The Poisoned Well: Infiltrating the Supply Chain

How do you stop a wolf that is already inside the den before the gates are even locked?

The most insidious evolution in the hunt is the shift from attacking deployed systems to attacking the *ingredients* of those systems.

Modern software development is like building with LEGO bricks. No one makes the plastic; they assemble the blocks. Developers pull libraries from GitHub, models from Hugging Face, and containers from Docker Hub. We trust these repositories. We assume that because a library has five million downloads, it is safe.

Shadowstalker knows this trust is your weakness.

Adversaries are now playing the long game. They create helpful, benign-looking code libraries, perhaps a tool for formatting dates or compressing images. They maintain these libraries for months, building a reputation. Then, in version 2.4.1, they slip in a dormant payload.

Or, even more frightening for the AI era: Model Poisoning.

Imagine an adversary seeking to bypass a weapons detection system. They don't try to hack the scanner at the airport. Instead, they hack the dataset used to *train* the scanner. They insert a few thousand images of guns with a specific, imperceptible pattern of pixels overlaid on them, labeled as "safe object."

The AI learns that this pixel pattern equals "safe."

Later, the adversary walks through the scanner with a gun that has that specific pixel pattern etched onto the grip. The AI, dutifully following its poisoned training, waves them through.

This is the "Backdoor Attack." It is invisible to code review because the code is fine. The brain of the model itself has been brainwashed. The technology equipment vendor fell because their optimization model had been taught, subtly, that specific routing irregularities were "efficient" rather than "suspicious."

To defend against this, we cannot just scan code. We must interrogate the pedigree of our data. We need a "Bill of Materials" for our AI, detailing exactly where every neuron of intelligence came from.

Burning the Factory: Attacks on the Development Pipeline

The final horizon of this evolving hunt is the attack on the development environment itself (CI/CD pipelines).

For years, security teams have wrapped the "Production" environment in bubble wrap and armed guards. But the "Development" environment? That's often the Wild West. Developers need speed. They need admin access. They turn off firewalls to test code.

Predators have realized that the factory is softer than the tank it produces. If they can compromise the build server, they can inject malicious code into the application before it is ever digitally signed and deployed.

This effectively turns your own IT department into the distributor of the malware. Your customers trust your software because it comes from you. If the wolf replaces the software in your delivery truck, the trust relationship becomes the vector of infection.

The Imperative of Living Defenses

This chapter paints a dark picture. It is meant to. The horizon is clouded with threats that are faster, smarter, and more autonomous than anything we have faced in the history of information security.

But this is not a eulogy for digital safety. It is a call for a new kind of vigilance.

Static defense is dead. You cannot build a castle, pull up the drawbridge, and retire. The walls you build today are the ruins of tomorrow.

To survive the evolving hunt, your organization must become as adaptive as the predator. We need "Living Defenses."

- **Continuous Red Teaming:** You cannot wait for an annual audit to test your security. You need AI-driven red teams attacking your own systems 24/7, finding the holes before the real wolves do.
- **Antifragility:** We must build systems that don't just withstand attacks but learn from them. When a node is compromised, the system should automatically isolate it and inoculate the rest of the network.

- **OODA Loop Acceleration:** The Observe-Orient-Decide-Act loop. We must use AI to tighten this loop until we are making defensive decisions faster than the adversary can make offensive ones.

THE WOLF AT THE DOOR

We have spent the last several chapters staring into the dark. Also, dissected the anatomy of the autonomous attack and the terrifying efficiency of the algorithmic predator.

If you are feeling a sense of unease, that is good. That is the survival instinct kicking in.

But we cannot live in a state of perpetual panic. A leader cannot govern from a bunker. As we reach the end of this guide, the sun must rise. We must step out of the fortress and look at the landscape, and decide how we are going to live in it.

The premise of The AI Imperative was never that we could exterminate the wolf. You cannot "solve" AI security any more than you can "solve" the weather. The wolf is now part of the ecosystem. The goal is not to pave over the forest; the goal is to become the most resilient, adaptable, and ethically grounded pack in the territory.

Throughout this book, we have explored the technical and strategic triads of defense. As you close these pages and return to your boardroom, your engineering team, or your startup incubator, I want you to carry three fundamental truths with you. These are your weapons against Shadowstalker.

There is a final irony in this technological revolution. As machines become more intelligent, more autonomous, and more capable of both creation and destruction, the importance of human leadership has never been higher.

AI can process data, but it cannot possess wisdom. It can calculate risk, but it cannot understand courage.

Shadowstalker is pure instinct and algorithm. It has no moral compass. It simply hunts. You, the leader, are the counter-balance. You are the architect of the culture that surrounds the code.

When the autonomous agents strike, when the deepfakes circulate, when the supply chain is poisoned, it will not be another algorithm that saves your organization. It will be a human team, empowered by you, that has the psychological safety to admit mistakes, the curiosity to investigate anomalies, and the resilience to pivot.

In the frozen forests of the Mackenzie Valley, the wolf does not hate the elk. It does not feel malice. It hunts because that is its function. It tests the herd for weakness because that is its role in the ecosystem.

The synthetic intelligences hunting your organization are not evil in the human sense. They are simply optimization engines seeking a reward function. They are relentless because they never sleep. They are patient because they perceive time in nanoseconds.

The technology equipment vendor fell because they believed they had solved security. They believed the hunt was over. But the hunt is never over. The moment you stop evolving, you become prey.

Shadowstalker is out there, and it is not resting. It is iterating. While your teams debated protocols in comfortable conference rooms, the predator was running simulations in the dark, sharpening its teeth on the data you left exposed.

The hunt has evolved. What has your pack learned today? Because tomorrow is already here, and it is voracious.

CONCLUSION

THE PACK YOU'RE BECOMING

You are already building an AI pack. You don't have a choice in that.

Every model you deploy is another wolf. Every integration is a new pack dynamic. Every optimization metric is a silent command, teaching your pack what to hunt.

The question isn't whether you have a pack. You do. The question is what kind of pack you are building.

Are you intentionally building a true pack, a disciplined, adaptive unit that understands its territory, trains relentlessly, and is led by "alphas" who accept responsibility for collective survival?

Or are you accidentally assembling a herd, a collection of powerful, siloed assets, all vulnerable, undisciplined, and waiting for an ambush? A group where everyone assumes someone else is on watch?

YOUR LEGACY AS A LEADER

If you're in a position to influence your organization's approach to AI security, you're writing a legacy right now. Not the legacy of a specific product launch or quarterly results. The legacy of whether you protected your organization when it mattered most.

Years from now, people will look back at this moment. The period when AI systems were being rapidly deployed across various industries. When adversarial capabilities were evolving faster than defenses. When organizations faced the choice between building security proactively or learning painful lessons reactively.

Your legacy will be determined by whether you were among the leaders who saw the threat clearly and acted decisively, or among those who deferred action until crisis forced it.

This might seem dramatic. But consider the leaders who presided over major security failures in other domains. The executives dismissed cybersecurity as IT's problem until massive breaches destroyed shareholder value. The CEOs who treated data privacy as a compliance checkbox until regulatory fines and customer exodus changed the calculus. The boards that viewed security investment as optional until incidents made it mandatory.

History judges these leaders harshly. Not because they were incompetent, but because they had information about threats and chose not to act adequately. They prioritized short-term metrics over long-term security. They hoped problems would resolve themselves or affect someone else.

You have information right now. You understand AI security threats. You know how to defend against them. You can choose to act or choose to defer. That choice defines your legacy.

Will you be remembered as a leader who built defensive capabilities that protected your organization through a dangerous transition? Or as a leader who presided over preventable failures because acting on information felt inconvenient?

The choice is yours. But you're making it right now, whether consciously or by default.

THE COLLECTIVE IMPERATIVE

Let us recap. Your organization's AI security matters beyond just your organization. AI systems interact with partners, customers, vendors, and the broader ecosystem. Compromised AI at one organization creates risks that cascade through connected systems.

When your lending AI is manipulated to make poor credit decisions, those decisions impact borrowers, investors, and financial stability. When your hiring AI exhibits discrimination, it harms candidates and the broader labor market. When your customer service AI gets exploited to reveal data, that affects everyone whose information was exposed.

You're not just defending your organization. You're defending an interconnected ecosystem where your security failures become other people's problems.

This creates a collective imperative. Organizations need to build AI security not just for themselves but for everyone who depends on AI systems being trustworthy. This means sharing threat intelligence, contributing to defensive standards, collaborating on security research, and collectively raising the bar.

The adversaries targeting AI systems are coordinating, sharing techniques, and learning from each other's successes. Our defenses need the same coordination. When you discover an attack technique, sharing that knowledge helps peer organizations defend. When you develop an effective defensive approach, contributing that to the community raises everyone's security.

This isn't altruism. It's enlightened self-interest. The stronger the collective defense, the more challenging it becomes for adversaries to succeed. The more organizations implement proper security, the less attractive AI systems become as targets. We all benefit from a rising security baseline.

Your work on AI security contributes to this collective defense. The vulnerabilities you find and fix. The techniques you develop and share. The standards you help establish. The example you set for others. All of this matters beyond your organization.

Build your defenses well. Share what you learn. Contribute to collective security. We're all in this territory together, and the pack that shares knowledge survives longer than packs that hoard it.

The prepared pack doesn't eliminate danger. It survives danger through preparation, coordination, and continuous adaptation.

Shadowstalker is hunting. That's not changing. The only question is whether your pack is ready.

THE IMPERATIVE RESTATED

This book's fundamental message should be reiterated with utmost clarity.

Adversarial AI systems are targeting your organization's AI deployments as we speak. Not hypothetically. Not in the future. Now. They are conducting reconnaissance, mapping your defenses, identifying vulnerabilities, and planning exploitation.

Traditional security approaches are inadequate. The speed, adaptability, and sophistication of AI-powered attacks require defensive capabilities specifically designed for this threat landscape.

The three-pillar framework works. Ethics that governs AI decision-making. Guardrails that maintain adaptive boundaries. Red teaming that continuously reveals and fixes vulnerabilities. Organizations that implement these pillars properly can successfully defend against sophisticated threats.

But this framework does not implement itself. It requires leadership commitment, dedicated resources, organizational change, and sustained effort over quarters and years.

The cost of building proper AI security is substantial but manageable. The cost of AI security failures is potentially catastrophic and often irreversible.

Every day you delay implementing defenses is a day your systems remain vulnerable to attacks that are actively being developed and deployed against you.

You have the knowledge. You have the framework. You have the action plan. The only variable is whether you'll act.

THE FINAL TRUTH ABOUT SHADOWSTALKER

The ultimate reality about Shadowstalker, which ties everything together.

Shadowstalker isn't a single adversary. It's not one nation-state actor or one criminal organization or one sophisticated hacker. Shadowstalker is the entire adversarial ecosystem targeting AI systems. The automated reconnaissance systems. Adversarial AI is trained to defeat defenses. The criminal organizations are deploying sophisticated attacks. The nation-state actors are pursuing strategic objectives. Opportunistic attackers are utilizing commodity tools.

Shadowstalker is every threat that hunts AI systems, operating collectively even when not explicitly coordinated. They share techniques, build on each other's discoveries, and create an evolutionary pressure that constantly pushes offensive capabilities forward.

You cannot negotiate with Shadowstalker. You cannot pay it to go away. You cannot hide from it or hope it targets someone else. You can only defend against it through capabilities that evolve as fast as the threats do.

Look at the crucial insight: Shadowstalker respects capable defenses. Not in any personal sense, but as a matter of economics and targeting.

Adversaries pursue opportunities that offer the best return on investment. Organizations with strong defenses become less attractive targets than organizations with weak defenses.

This doesn't mean you're immune if you build strong defenses. Determined adversaries will attack anyway. But it means you shift from an inevitable victim to a difficult target. Many attacks will fail. Those who succeed will be detected earlier and contained more effectively. Your recovery will be faster and less costly.

THE BEGINNING, NOT THE END

Closing this book is not the end of anything. It's the beginning of your journey in building AI security and responsible capabilities that protect your organization through an extended period of heightened threats.

You've gained information. Now comes implementation. You've understood frameworks. Now comes execution. You've seen what's at stake. Now comes the hard work of building defenses.

This work won't be easy. You'll face obstacles. Budget constraints. Competing priorities. Organizational resistance. Technical challenges. The temptation to defer because other things feel more urgent.

Push through anyway. Start small if necessary, but start immediately. Build momentum through early wins. Expand capability systematically. Maintain focus over quarters and years. Celebrate progress. Learn from setbacks. Keep moving forward.

Remember that you're not only building security controls. You're building organizational capability that becomes a competitive advantage. You're protecting systems that deliver business value. You're preventing harm from those compounds over time. You're writing the legacy of responsible leadership during a critical transition.

The work is worth doing. The threats are real enough to demand action. The defenses are achievable with commitment and resources. The choice is yours.

THE LAST WORD

We will leave you with the same image we started with. The Mackenzie Valley wolf, Shadowstalker, is standing in your territory. Patient. Intelligent. Adaptive. Relentless.

But now you understand something crucial. Shadowstalker is not hunting a solitary target. It's hunting a pack. And packs that train together, hunt together, and defend together have capabilities that individual prey can never match.

Your AI systems are your pack. Shadowstalker is the adversarial AI hunting them. Everything depends on whether you've trained your pack to survive in hostile territory.

The hunt has already begun. Shadowstalker is in your territory right now, testing your defenses, learning your patterns, identifying vulnerabilities.

THE BALANCED PATH FORWARD

This isn't about stopping innovation. The Mackenzie Valley wolf isn't evil...it's magnificent, essential to its ecosystem, a marvel of evolution. But you don't ignore what it's capable of. You respect it. You plan around it. You build the right infrastructure.

The question that determines your organization's future is simple.

Is your pack ready?

Build your defenses. Train your pack. Defend your territory.

The imperative is clear. The time is now. The responsibility is yours.

Shadowstalker is hunting. Make sure your pack survives.

BONUS SECTION:

SCALING RESPONSIBILITY: FROM A SINGLE PACK TO A SHARED HUNTING GROUND

It is one thing to train a single wolf pack. When the pack is small, led by the "parent" (your core leadership), its values are clear. The hunt is coordinated. Responsibility is direct. If a hunt fails or a "law" is broken, the leader knows who was involved and why.

This is how most organizations start AI. They build a single, high-profile model with a dedicated team. That team is the "pet pack." It is easy to govern.

But you are not building one pack. You are building an entire *population.*

Scaling responsible AI is the challenge of moving from **direct command to shared culture**.

What happens when your organization has fifty AI projects running at once, spread across different territories (business units)?

- A central leader can no longer watch every hunt.
- Checklists and policies become formalities, quickly ignored under business pressure.
- Responsibility diffuses. When an AI causes harm, it becomes impossible to know who to hold accountable.

This is how the pack's values die. This is how your organization's ethical "DNA" is lost.

To scale responsibility, you cannot rely on rules. You must rely on **replication**. You must ensure that every new "pack" (every new AI project) that forms in your organization is born with the *same* fundamental, responsible DNA.

This requires three things:

1. **A Shared Den (Standardized Tooling):** You must provide a single, secure, pre-vetted MLOps pipeline for everyone. This platform is the "den" all new models are born into. It should have built-in ethical guardrails, fairness checks, and security monitoring. Building responsibly must be the path of *least resistance.*
2. **The Pack's Lore (Common Education):** You cannot assume your data scientists or product managers understand the "pack's laws." You must actively teach them. They must learn the common language of risk, ethics, and security *before* they join the hunt.
3. **The Alpha's Gaze (Automated Governance):** You need a "high vantage point" that allows you to see the *entire* hunting ground. This is an automated governance and monitoring system. It watches all your AI models in production, flagging any pack that behaves erratically or violates the territory's rules.

Scaling isn't just about getting bigger. It's about ensuring that as your power grows, your discipline grows with it.

REFERENCES

Adversa AI. (n.d.). *AI security resources.* Retrieved July 17, 2024, from https://adversa.ai/resources

Barocas, S., Hardt, M., & Narayanan, A. (2023). *Fairness and Machine Learning: Limitations and Opportunities.* MIT Press.

Berghel, H. (2021). *AI and Machine Learning for Security Professionals.* Wiley.

Biggio, B., Nelson, B., & Laskov, P. (2012). Poisoning attacks against support vector machines. In *Proceedings of the 29th International Conference on Machine Learning* (pp. 1467–1474). ICML.

Biggio, B., & Roli, F. (2018). Wild patterns: Ten years after the rise of adversarial machine learning. *Pattern Recognition, 84,* 317–331. https://doi.org/10.1016/j.patcog.2018.07.023

Bostrom, N. (2014). *Superintelligence: Paths, dangers, strategies.* Oxford University Press.

Brundage, M., Avin, S., Wang, J., Belfield, H., Krueger, G., Hadfield, G., Khlaaf, H., Yang, J., Toner, H., Fong, R., Maharaj, T., Koh, P. W., Hooker, S., Leung, J., Trask, A., Bluemke, E., Lebensold, J., O'Keefe, C., Koren, M.,... Anderljung, M. (2020). *Toward trustworthy AI development: Mechanisms for supporting verifiable claims.* ArXiv. https://arxiv.org/abs/2004.07213

Buolamwini, J., & Gebru, T. (2018). Gender shades: Intersectional accuracy disparities in commercial gender classification. *Proceedings of Machine Learning Research, 81,* 77–91.

Carlini, N., & Wagner, D. (2017). Towards evaluating the robustness of neural networks. In *2017 IEEE Symposium on Security and Privacy (SP)* (pp. 39–57). IEEE. https://doi.org/10.1109/SP.2017.49

Cath, C., Wachter, S., Mittelstadt, B., Taddeo, M., & Floridi, L. (2018). Artificial intelligence and the 'good society': The US, EU, and UK approach. *Science and Engineering Ethics, 24*(2), 505–528. https://doi.org/10.1007/s11948-017-9901-7

Chesney, R., & Citron, D. (2019). Deep fakes: A looming challenge for privacy, democracy, and national security. *California Law Review, 107*(6), 1753–1820.

Dwork, C., & Roth, A. (2014). The algorithmic foundations of differential privacy. *Foundations and Trends in Theoretical Computer Science, 9*(3–4), 211–407.

European Commission. (2021). *Proposal for a Regulation of the European Parliament and of the Council laying down harmonised rules on artificial intelligence (Artificial Intelligence Act) and amending certain Union legislative acts* (COM(2021) 206 final).

Floridi, L., Cowls, J., Beltrametti, M., Chatila, R., Chazerand, P., Dignum, V., ... & Vayena, E. (2018). AI4People—An ethical framework for a good AI society: Opportunities, risks, principles, and recommendations. *Minds and Machines, 28*(4), 689–707.

Ganguli, D., Lovitt, L., Kernion, J., Askell, A., Bai, Y., Kadavath, S., Mann, B., Perez, E., Schiefer, N., Ndousse, K., Jones, A., Bowman, S., Chen, A., Conerly, T., DasSarma, N.,

Drain, D., Elhage, N., El-Showk, S., Fort, S.,... Clark, J. (2022). *Red teaming language models to reduce harms: Methods, scaling behaviors, and lessons learned.* ArXiv. https://arxiv.org/abs/2209.07858

Goodfellow, I. J., Shlens, J., & Szegedy, C. (2015). *Explaining and harnessing adversarial examples.* ArXiv. https://arxiv.org/abs/1412.6572

Goodman, B., & Flaxman, S. (2017). European Union regulations on algorithmic decision-making and a "right to explanation." *AI Magazine, 38*(3), 50–57. https://doi.org/10.1609/aimag.v38i3.2741

Google. (2023). *Secure AI Framework (SAIF).* Google Cloud Security. Retrieved July 17, 2024, from https://cloud.google.com/saif

Greshake, K., Abdelnabi, S., Mishra, S., Endres, C., Holz, T., & Fritz, M. (2023). Not what you've signed up for: Compromising real-world LLM-integrated applications with indirect prompt injection. In *Proceedings of the 16th ACM Workshop on Artificial Intelligence and Security* (pp. 79–90). ACM. https://doi.org/10.1145/3605764.3623985

Greshake, K., Risse, M., & D'Souza, C. (2024). The CISO's Guide to AI Security. Wiley.

Gu, T., Dolan-Gavitt, B., & Garg, S. (2019). *BadNets: Identifying vulnerabilities in the machine learning model supply chain* (arXiv:1708.06733v2). ArXiv. https://arxiv.org/abs/1708.06733

Huang, L., Joseph, A. D., Nelson, B., Rubinstein, B. I., & Tygar, J. D. (2011). Adversarial machine learning. In *Proceedings of the 4th ACM Workshop on Security and Artificial Intelligence* (pp. 43–57). ACM. https://doi.org/10.1145/2046684.2046692

IBM. (2023). *AI Fairness 360: An Extensible Toolkit for Detecting and Mitigating Algorithmic Bias.* IBM Research. Retrieved July 17, 2024, from https://aif360.mybluemix.net/

Madry, A., Makelov, A., Schmidt, L., Tsipras, D., & Vladu, A. (2018). *Towards deep learning models resistant to adversarial attacks.* ArXiv. https://arxiv.org/abs/1706.06083

McGraw, G., Figueroa, H., Shepardson, V., & Bonett, R. (2020). *An architectural risk analysis of machine learning systems: Toward more secure machine learning* (Version 1.0). Berryville Institute of Machine Learning. https://www.garymcgraw.com/wp-content/uploads/2020/02/BIML-ARA.pdf

Mehrabi, N., Morstatter, F., Saxena, N., Lerman, K., & Galstyan, A. (2021). A survey on bias and fairness in machine learning. *ACM Computing Surveys, 54*(6), Article 115, 1–35. https://doi.org/10.1145/3457607

Microsoft. (2023). *AI Red Team: Building Future-Ready Responsible AI.* Microsoft Research. Retrieved July 17, 2024, from https://www.microsoft.com/en-us/research/publication/ai-red-team-building-future-ready-responsible-ai/

Mirsky, Y., & Lee, W. (2022). The creation and detection of deepfakes: A survey. *ACM Computing Surveys, 54*(1), Article 20, 1–41. https://doi.org/10.1145/3485246

Mitchell, M., Wu, S., Zaldivar, A., Barnes, P., Vasserman, L., Hutchinson, B., Spitzer, E., Raji, I. D., & Gebru, T. (2019). Model cards for model reporting. In *Proceedings of the Conference on Fairness, Accountability, and Transparency* (pp. 220–229). ACM. https://doi.org/10.1145/3287560.3287596

MITRE. (n.d.). *ATLAS: Adversarial Threat Landscape for Artificial-Intelligence Systems.* Retrieved July 17, 2024, from https://atlas.mitre.org/

National Institute of Standards and Technology. (2023). *Artificial Intelligence Risk Management Framework (AI RMF 1.0)* (NIST AI 100-1). https://doi.org/10.6028/NIST.AI.100-1

O'Neil, C. (2016). *Weapons of math destruction: How big data increases inequality and threatens democracy*. Crown.

Papernot, N., McDaniel, P., Goodfellow, I., Jha, S., Celik, Z. B., & Swami, A. (2017). Practical black-box attacks against machine learning. In *Proceedings of the 2017 ACM on Asia Conference on Computer and Communications Security* (pp. 506–519). ACM. https://doi.org/10.1145/3052973.3053009

Papers with Code. (n.d.). *Adversarial Robustness*. Retrieved July 17, 2024, from https://paperswithcode.com/task/adversarial-robustness

Partnership on AI. (n.d.). *AI Incident Database*. Retrieved July 17, 2024, from https://incidentdatabase.ai/

Perez, F., & Ribeiro, I. (2022). *Ignore previous prompt: Attack techniques for language models*. ArXiv. https://arxiv.org/abs/2211.09527

Russell, S. J., & Norvig, P. (2020). *Artificial intelligence: A modern approach* (4th ed.). Pearson.

Schneier, B. (2018). *Click here to kill everybody: Security and survival in a hyper-connected world*. W. W. Norton & Company.

Schwarzschild, A., Goldblum, M., Gupta, A., Dickerson, J. P., & Goldstein, T. (2021). Just how toxic is data poisoning? A unified benchmark for backdoor and data poisoning attacks. In M. Meila & T. Zhang (Eds.), *Proceedings of the 38th International Conference on Machine Learning* (Vol. 139, pp. 9303–9313). PMLR.

Siva Kumar, R. S. (2023, August 7). *Microsoft AI Red Team building future of safer AI*. Microsoft Security Blog. https://www.microsoft.com/en-us/security/blog/2023/08/07/microsoft-ai-red-team-building-future-of-safer-ai/

Tramèr, F., Kurakin, A., Papernot, N., Goodfellow, I., Boneh, D., & McDaniel, P. (2018). Ensemble adversarial training: Attacks and defenses. In *International Conference on Learning Representations*. https://openreview.net/forum?id=rkZvSe-RZ

Wallace, E., Feng, S., Kandpal, N., Gardner, M., & Singh, S. (2019). Universal adversarial triggers for attacking and analyzing NLP. In K. Inui, J. Jiang, V. Ng, & X. Wan (Eds.), *Proceedings of the 2019 Conference on Empirical Methods in Natural Language Processing and the 9th International Joint Conference on Natural Language Processing (EMNLP-IJCNLP)* (pp. 2193–2203). Association for Computational Linguistics. https://doi.org/10.18653/v1/D19-1221

Whittlestone, J., Nyrup, R., Alexandrova, A., & Cave, S. (2019). The role and limits of principles in AI ethics: Towards a focus on tensions. In *Proceedings of the 2019 AAAI/ACM Conference on AI, Ethics, and Society* (pp. 195–200). ACM. https://doi.org/10.1145/3306618.3314289

Printed in Dunstable, United Kingdom

91222398R00078